# GUIDE to the IDENTIFICATION

of some of the more

DIFFICULT VASCULAR PLANT SPECIES

with particular application to the

Watsonian vice-counties 66-70

Durham, Northumbria, and Cumbria

M.J. Wigginton

G.G. Graham

**Nature Conservancy Council**

**New address**
The headquarters of the Nature Conservancy Council are now at Northminster House Peterborough PE1 1UA.
Catalogues of current publications are also available from this address.

57601

**January 1981** ISBN 0 86139 133 0

# Preface

This guide to some of the more difficult groups of vascular plants was intended primarily for recorders in the north of England, but much of the guide is applicable to the whole country. In compiling this booklet, we have attempted to bring together the results of taxonomic work published since the appearance of the Flora of the British Isles, (Clapham, Tutin, and Warburg, 1962).
However, it is important to point out that our intention was not to produce a Flora, but merely to provide new keys and descriptions to assist in the identification of plants.
We have made full use of papers that have appeared in Watsonia, and other journals, and also of a wide range of standard works, the most important of which are,

Flora Europaea, Vols 1-5, T.G. Tutin Ed. Cambridge University Press, 1964-80

A Field Guide to the Trees of Britain and Northern Europe, by A. Mitchell. Collins, 1974

Grasses, by C.E. Hubbard. 2nd Edn. Penguin, 1968

British Sedges, by A.C. Jermy & T.G. Tutin B.S.B.I. 1968

Atlas of Ferns of the British Isles, A.C. Jermy Ed. B.S.B.I. 1978

Hybridisation and the Flora of the British Isles, C.A. Stace Ed. Cambridge University Press, 1975

Drawings of British Plants, by S. Ross-Craig. Bell & Hyman

In addition, specialists have provided valuable information on a number of groups.

We cover fully, or in part, 148 genera, the arrangement following that of Clapham, Tutin, and Warburg, 2nd Edn. 1962. Nomenclature mainly follows that of Flora Europaea Vols 1-5, but in few instances, more recent work has necessitated the use of the name currently accepted. The keys and descriptions supplement, or mainly update those in CTW, 1962.
In the text, where reference is made to 'the region' or 'our area', this is understood to refer to the area comprising the Watsonian vice-counties 66 - 70, i.e. Durham, Northumbria, and Cumbria. Sources of information are given at the end of the species or genus accounts.

In a number of genera (e.g. Rosa, Rubus, Hieracium, Euphrasia) determination is a matter for the specialist, but keys are provided to encourage the wider study of these critical groups. It should be emphasised that it is important to check determinations (from keys) by reference to full descriptions which appear in the standard texts.

## Acknowledgements

We are especially indebted to Dr G. Halliday for his many valuable comments on much of the text.
Thanks are also due to a number of specialists for their help in some of the more critical genera. They are all mentioned in the main text, but in particular we acknowledge our debt to Dr A.J. Silverside for the key to Mimulus, and for his helpful comments on several other groups. Thanks are due to Dr N.T.H. Holmes (Ranunculus), Dr M.G. Daker (Fumaria), A. Newton (Rubus), Dr M.E. Bradshaw (Alchemilla), Dr P.F. Yeo (Euphrasia), and D. Simpson (Elodea), among others.

We should also like to thank Miss S. Collins for preparing almost all the typescript.

We are grateful to a number of publishing houses for permission to reproduce drawings. Those included in this guide are reproduced from the following

Illustrations of British Plants, by S. Ross-Craig, by permission of the publishers, Bell & Hyman Ltd. These appear on pages 16-19, 21, 22, 24-30, 35, 38, 46, 50, 55, 59-61, 65, 77, 86, 94, 99, 101, 112, 114, 115, 117, 118, 122, 124.

Flora of the British Isles, by Clapham, Tutin, and Warburg, 2nd Edn, 1962, by permission of the publishers, Cambridge University Press. These appear on page 58.

Welsh Ferns, by H.A. Hyde, A.E. Wade, and S.G. Harrison, 4th Edn, 1969, by permission of the publisher, the National Museum of Wales. These appear on pages 4, 6, 8.

A Field Guide to the Trees of Britain and Northern Europe, by A. Mitchell, 1974, by permission of the publishers, Collins Ltd. These appear on page 71.

---

The authors, M.J.W. and G.G.G., are solely responsible for what appears in this guide. Comments and suggestions for the improvement of keys would be welcomed.

M.J. Wigginton

G.G. Graham

January 1981

---

M.J. Wigginton, Nature Conservancy Council, Calthorpe House, Calthorpe Street, Banbury, Oxon.

G.G. Graham, The Vicarage, Hunwick, near Crook, Durham.

<u>ISOETES</u> L.

Inadequate or inaccurate descriptions of I. setacea in British Floras have hitherto made its identification and differentiation from I. lacustris more difficult.
Leaf-shape is the best macro-character for separating the two species, but not when the plant is dried. Megaspore ornamentation is the most reliable overall character, but is not a field character.

Leaf tapering gradually and evenly throughout its length to a long, hair-like tip, grass-green, supple. Megaspores covered with long, sharp, fragile spines.
I. setacea Lam. (I. echinospora Durieu)

Leaf not tapering for most of its length, but narrowing about 5mm from the tip to an asymmetric point, usually much darker green, stiff and brittle. Megaspores covered with short, blunt tubercles. I. lacustris L.

I. lacustris is usually larger, but there is a considerable overlap. Leaf posture and number have little significance. Leaves of I. setacea are stiff at the base, but adhere like a brush when taken out of the water. However, young leaves of I. lacustris may also adhere, and old leaves of I. setacea may not, but otherwise this is a useful differential character.

R. Stokoe, Watsonia, 12(1), 51-52 (1978)

EQUISETUM x LITORALE Kuhlew. ex Rupr. (E. arvense x E. fluviatile)

Hybrids are usually intermediate between the parents in morphology. They can be distinguished from E. fluviatile by the smaller stem hollow, branches with fewer ribs, sheaths more campanulate than cylindrical; from E. arvense by the larger stem hollow, and glabrous rhizomes. They are, however, often very variable and sometimes closely resemble either parent (e.g. in drier habitats they may approach E. arvense, and in wetter habitats E. fluviatile).

In recent years the hybrid has been found in a number of localities in Cumbria, and is probably widespread in the region.

J.G. Duckett & C.N. Page, in Stace (1975)

Stem transverse section x8

E. arvense E.x litorale E. fluviatile

EQUISETUM VARIEGATUM Schleich. ex Web. & Mohr.

Care should be taken to distinguish prostrate, unbranched (or sparingly-branched) forms of E. palustre L. from E. variegatum Schleich. ex Web. & Mohr.

| | variegatum | palustre |
|---|---|---|
| stem grooves | 4-10, moderately deep | 4-8, deep |
| sheaths | 2-4mm, green with a black band around the top, rather loose | 4-12mm, green, loose |
| teeth | 4-10, triangular-ovate to triangular-lanceolate, at first subulate, the tip caducous, leaving an obtuse apex | 4-8, triangular-subulate, tip not caducous |
| cone | short, 5-7mm | longer, 10-30mm |
| apical sporang--iophore | prominently apiculate | obtuse |

Equisetum variegatum is a scarce plant of dune-slacks, and basic flushes in upland areas.

OPHIOGLOSSUM AZORICUM C.Presl.
(O. vulgatum L. subsp. ambiguum (Coss. & Germ.) E.F. Warb.)

This taxon requires further investigation as its relationship to O. vulgatum L. is not clear.

Plants having 14 or fewer sporangia, a sterile blade less than 3.5cm, and occurring in short turf near the sea are named O. azoricum for the time being.

Atlas of Ferns (1978)

HYMENOPHYLLUM Sm.

The two British species are readily distinguished by indusium and leaf characters. H. tunbrigense is rare in our area and restricted to a few sheltered ravines.

Indusium flattened, valves orbicular, with a wide, irregularly and sharply-toothed mouth. Frond ± flat. H. tunbrigense (L.) Sm.

Indusium not flattened, valves ovate, entire. Pinnae narrower, bent back from the rachis. H. wilsonii Hook.

POLYPODIUM VULGARE complex

The leaf shape, often given as a useful diagnostic character, is not reliable and is at best only an indicator. The most reliable diagnostic characters appear to be the following:

i) the colour of the annulus cells

ii) the number and width of the basal cells i.e. the cells separating the sporangium-stalk from the annulus

iii) the shape of the rhizome-scales

Microscopic examination is needed for confirmation.

P. vulgare L.

Annulus cells (7)10-14(17), dark reddish-brown, and contrasting sharply with the rest of the sporangium. Basal cells usually 1, as wide as, or very little wider than the annulus. No large branched paraphyses, but small glandular hairs often present. Rhizome scales normally 3-6mm, usually more triangular in shape than those of interjectum. Leaves normally linear-lanceolate (i.e. more than half the pinnae ± the same length) but the leaf shape is not reliable. New leaves produced in early summer, spores ripening in July-August. Stomatal guard cells 43-58µm. Spores less than 70µm. Usually associated with acid habitats, and also as an epiphyte. Widespread.

P. interjectum Shivas

Annulus cells (3)6-10(13), usually pale golden-brown, but they do vary from pale buff to pale or bright yellow, or even orange-yellow. This variability may in itself be a useful diagnostic character. Basal cells usually 2-3, much wider than the annulus. No large branched paraphyses, but small glandular hairs often present. Rhizome-scales 3.5-11mm, often rather abruptly contracted above the dilated and relatively broader base. Leaves ovate to ovate-lanceolate (i.e. the longest pinnae 4th to 6th pair from the base) but leaf shape is not reliable. Stomatal guard cells 58-71µm. Spores more than 74µm. New leaves produced in summer, spores ripening July-September. Usually in basic habitats, especially walls. Widespread.

P. australe Fée

Annulus cells (2)4-10(18), nearly always bright yellow. Basal cells usually 3-4, much wider than annulus. Paraphyses large and branched, averaging 970 µm. long. Rhizome scales 5-16mm, mostly linear-lanceolate. Leaves broadly-ovate to triangular. New leaves produced in the autumn, spores ripening in late autumn or the following spring. Stomatal guard cells 43-65 mµlong. Spores less than 74µm long. On limestone, very local.

Drawings of annulus and paraphyses below.

Paraphyses of

a,b P.australe x P.interjectum

c-e P.australe

f-j P.interjectum

( x950 )

Annulus cells of
P. australe
x200

R.H. Roberts, Watsonia 8(2) 121-134 (1970)

Polypodium x mantoniae (Rothm.) Shivas
(P. interjectum x P. vulgare)

Polypodium x font-queri Rothm.
(P. australe x P. vulgare)

ASPLENIUM L.

A. billotii F.W. Schultz (A. obovatum auct.) is rare in the region, with only one locality known at present (v.c.70). However, it is perhaps overlooked or mistaken for an immature plant of A. adiantum-nigrum L. They are distinct in leaf-shape and sori-shape.

Lamina lanceolate, i.e. basal pinnae not the longest; sori oblong.
A. billotii F.W. Schultz

Lamina triangular, i.e. the basal pair of pinnae the longest; sori proportionately longer.
A. adiantum-nigrum L.

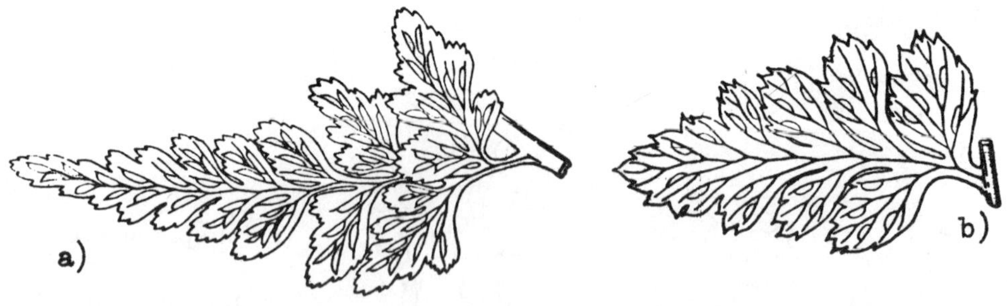


Median pinna from average-sized frond a) A. adiantum-nigrum (x2)
b) A. billotii (x2)

ASPLENIUM TRICHOMANES complex

Two subspecies are recognised in Britain, the distinctions between the two are most apparent in the upper part of the leaf. The most evident difference in the field is the conspicuously concave upper pinnae in subsp. trichomanes; in subsp. quadrivalens this is convex with the margins rolled (or, less commonly, flat). Both subspecies are very plastic, and appear atypical when growing in exposed as opposed to sheltered conditions.

Plants of sheltered sites

subsp. trichomanes
Rachis thin, wiry, red-brown. Pinnae distant, mostly alternate, obliquely-inserted, with a distinct stalk, asymmetric, to 8mm long, often perceptibly auriculate. Lamina delicate.
Sori to 2mm, relatively few, 4-6(9). Indusia narrow and delicate.
Calcifuge.

subsp. quadrivalens D.E. Meyer emend Lovis
Rachis thick, often dark-brown or blackish-brown. Pinnae more crowded, mostly opposite, with approximately transverse insertion, almost sessile, symmetrical, usually oblong, rarely auriculate, to 11mm long. Lamina thicker, with often crowded, more numerous sori, 4-9(12), to 3mm long. Indusia conspicuous. Predominantly calcicole.

Plants of exposed sites

Both subspecies are much reduced in size. Subsp. trichomanes is still distinguishable by its delicate rachis, and orbicular, flat or concave pinnae, with small, relatively few sori.
Subsp. quadrivalens has a stout rachis and oblong pinnae, with crowded, more numerous sori.

Micro-characters may also be used, but in most characters there is a considerable overlap. Where measurements are taken, mean values need to be obtained.

subsp. trichomanes
Rhizome scales to 3.5mm long, with a central red-brown stripe. Spores paler than those of subsp. quadrivalens, mean length (23)29-36(46) µm.
Mean length of stomatal guard cells (31)38-43(52) µm.

subsp. quadrivalens
Rhizome scales without a red brown stripe, to 5mm. Spores darker than those of subsp. trichomanes, mean length (27)34-43(50) µm. Mean length of stomatal guard cells (35)41-49(57) µm.

J.D. Lovis, British Fern Gazette 9 147-160 (1964); Fern Atlas (1978)

Asplenium x murbeckii Dorfl. (A. ruta-muraria x A. septentrionale)

Asplenium x alternifolium Wulfen. (A. septentrionale x A. trichomanes subsp. trichomanes

POLYSTICHUM L.

The two species P. aculeatum and P. setiferum are sometimes confused. P. aculeatum can be detected in the field by its stiff appearance, with a darker-green, somewhat coriaceous lamina. By contrast, P. setiferum is usually somewhat drooping or arching.

Leaves rigid, somewhat coriaceous, dark-green; basal pinnae much smaller than those in the middle of the frond, and only pinnatifid; margins at the base of the pinnules meeting each other usually at an acute angle; pinnules not stalked; spores papillate, mean size 41.3µm. P. aculeatum (L.) Roth.
(P. lobatum (Huds.) Chevall.)

Leaves ± flaccid, not coriaceous, rather paler in colour; basal pinnae long, spreading, and pinnate; margins at the base of the acroscopic pinnules meeting each other usually at about a right angle; pinnules stalked; mean spore size 30.5µm, with a winged perispore. Scarce in the Eastern parts of the region. P. setiferum (Forsk.) Woyner

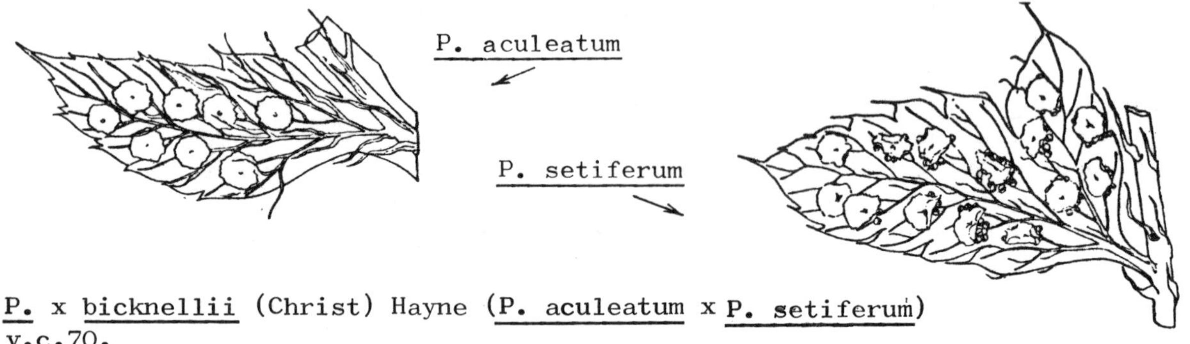


P. x bicknellii (Christ) Hayne (P. aculeatum x P. setiferum)
v.c.70.

DRYOPTERIS FILIX-MAS complex

Species of this complex have been much confused, and identification must always be on a combination of characters.

1. Lamina texture thick, somewhat glossy and dark-green on upper-side (yellow-green when young); petiole and rachis bearing long, narrow scales with a dark base and usually a dark centre; junction of rachis and pinna-axis darkly-coloured; lower margins of pinnules parallel, with few teeth. Immature indusia thick, with involute margins. D. affinis (Lowe)Fraser-Jenkins
(D. pseudomas (Wollaston) Holub & Pouzar, D.borreri Newm.)

1. Lamina texture thin, mid- or grey-green on top; petiole and rachis bearing both narrow and wide concolorous scales; junction of rachis and pinna-axis not darkly-coloured; pinnule-shape tapering, lower margins usually bearing teeth or lobes. Immature indusia varions.

   2. Lamina grey-green; teeth of pinnules obtuse, spreading fan-like at the apex of the pinnule; immature indusia thick, margins involute. Screes, etc., in upland areas. D. oreades Fomin.(D.abbreviata (DC.)Newm.)

   2. Lamina mid-green; teeth of pinnules acute, converging towards the apex; immature indusia thin, margins spreading. D. filix-mas (L.) Schott

Dryopteris x tavelii Rothm.
(D. filix-mas x D.pseudomas) Probably widespread, and greatly under-recorded

Dryopteris x mantoniae Fraser-Jenkins & Corley
(D. filix-mas x D. oreades)

Fern Atlas (1978)

## DRYOPTERIS AUSTRIACA complex

| D. expansa (D. assimilis) | D. austriaca (D. dilatata) | D. carthusiana (D. lanceolatocristata) |
|---|---|---|
| **Rhizome** | | |
| Erect to semi-prostrate | Erect (rarely semi-prostrate) | Decumbent or creeping |
| **Leaves** | | |
| Forming an open, rigid shuttlecock. | Forming a dense shuttlecock. | More distant, not forming a distinct shuttlecock. |
| **Scales** | | |
| With dark median stripe, or concolorous. | With dark median stripe, or concolorous. | Concolorous. |
| **Petiole** | | |
| About 2/3 length of lamina. | About 2/3 length of lamina. | About equalling lamina; relatively thin. |
| **Lamina** | | |
| 7-60cm long, usually ovate, light-green, thin. | 30-150cm long, usually ovate, not narrowed at base, olive-green. | 30-120cm long, oblong-lanceolate to lanceolate, not narrowed at base, light- to yellowish-green. |
| **Sori** | | |
| Often relatively large, 0.5-1.5mm in diam. | 0.5-1mm in diam. | 0.5-1.5mm in diam. |
| **Spores** | | |
| Perispore thin, pale, with smaller, widely-spaced, acute spinules to 1µm. | Perispore, dark-brown with dense, rather irregular, obtuse spinules. | |
| **Occurrence** | | |
| Scarce in upland regions. | Woods, heaths. Common. | Damp woods, marshes, heaths; local in the east of the region, frequent in the west. |

It should be noted that plants of D. austriaca (Jacq.) Woynar with concolorous scales on the rachis cannot be distinguished from D. carthusiana (Villar) H.P. Fuchs on this character. D. austriaca sometimes produces stolons from the base of the petiole (which can be mistaken for the rhizomes of D. carthusiana. Leaves on such stoloniferous shoots have concolorous scales. The leaves eventually form a 'shuttlecock' and the stem continues growth as an erect rhizome.

In D. expansa (C. Presl.) Fraser-Jenkins & Jermy, the basal pinnule on the lower side of the lowest pinna is usually twice as long as the pinnule on the upper side. The lower pinnule in addition, is usually at least half as long as the lowest pinna itself. However, diagnosis should not be made on these characters alone, as they are somewhat variable.

Hybrids confirmed in the region:

*Dryopteris* x *brathaica* Fraser-Jenkins & Reichstein
(*D. carthusiana* x *D. filix-mas*)

*Dryopteris* x *ambrosiae* Fraser-Jenkins & Jermy
(*D. austriaca* x *D. expansa*)

*Dryopteris* x *deweveri* (Jansen) Jansen & Wachter
(*D. carthusiana* x *D. austriaca*)

## DRYOPTERIS VILLARII (Bell) Woynar ex Schinz & Thell subsp. SUBMONTANA Fraser-Jenkins & Jermy

*D. villarii* is a very local plant in Britain, almost confined to the limestone areas of the north-west. It is locally frequent in limestone areas in v.c.69. especially in the grikes of limestone pavement. The green lamina has a distinct bluish tinge, which often contrasts with other ferns such as *Dryopteris filix-mas* and *D. pseudomas* which also occur in the grikes. The lamina is ± densely-glandular on both sides with short-stalked glands. The pinnules have a very distinctive shape, and the teeth are not spinulose.

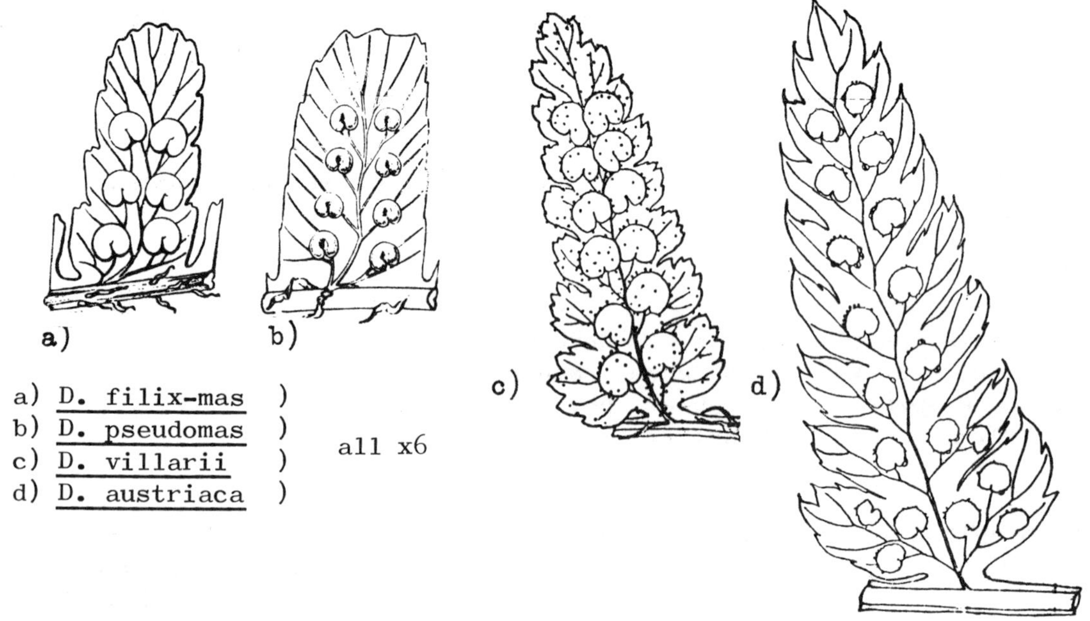


a) *D. filix-mas* )
b) *D. pseudomas* )
c) *D. villarii* ) all x6
d) *D. austriaca* )

## LARIX Mill

Species in this genus are often noted as '*L. decidua*' without distinguishing between true *L. decidua* and *L. kaempferi* (*L. leptolepis*) and their hybrid, all of which occur in plantations, shelter-belts etc.

Young shoots yellowish: leaves green, on short shoots 12-30 mm long, without grey stomatal furrows beneath; cones ovoid, the scales erect.
*L. decidua* Mill

Young shoots dark orange-red, pruinose; leaves glaucous, on short shoots 15-40mm long, with 2 broad, grey, stomatal furrows beneath; cones subglobose, the scales strongly reflexed. *L. kaempferi* (Lamb) Carr.
(*L. leptolepis* (Sieb. & Zucc.) Endl.)

The hybrid is widely planted. The leaves are intermediate in length and colour, and the cone-scales are strongly patent but not reflexed.

The hybrid is fertile and can backcross to both parents; mixed plantations of hybrids and parent species are not uncommon.

A. Mitchell (1974)

## JUNIPERUS COMMUNIS L.

Both subspecies occur in our area, but subsp. *communis* can be very dwarf and then mistaken for subsp. *nana*.

Prickly to the touch. Leaves spreading almost at right angles to the stem, 8-19 x 1 mm, contracted to a long point. Fruit globose. The lowland form.
subsp. *communis*

Scarcely prickly to the touch. Leaves ascending or loosely appressed, 4-10 x c.1.5 mm, more suddenly contracted to a shorter point. Rocks and moors on mountains. subsp. *nana* Syme

However, these represent two extremes, and many intermediates occur.

## RANUNCULUS REPTANS L.

This predominantly northern species is usually considered to be absent from Britain. It is characterised by linear leaves and arching stolons, which root freely at the nodes. Very similar plants occur by Ullswater (v.c.69 and 70) and remain distinct from *R. flammula* in cultivation. Superficially similar plants from elsewhere in Cumbria become more robust in cultivation with broader leaves and are referable to *R. flammula* subsp. *radicans*. The status of the Ullswater plants is being investigated further; it has been suggested that they are relict hybrid segregates of the two species.

RANUNCULUS FICARIA L.

Of the two fairly well-marked subspecies occurring in the region, subsp. bulbifer is certainly under-recorded.

R. ficaria L. subsp. ficaria (subsp. fertilis (Clapham) Lawalrée)

Axillary bulbils absent. Most of the achenes usually developing. Diploid. Pollen partly fertile. More generally in open places - pastures, hedgerows etc.

R. ficaria L. subsp. bulbifer (Marsden-Jones) Lawalrée

Bulbils present in the axils of the leaves; usually only a few achenes developing. Tetraploid. Pollen partly sterile. Plant more usually found in shady and wetter places. Large, yellowing mats in late spring (evidently spreading vegetatively) are good indicators of this species.

There is a considerable overlap between the subspecies in petal-size and shape, and flower-size.

RANUNCULUS L. Sect. BATRACHIUM (DC.) Gray

This group presents the recorder with considerable difficulties, mainly because of the considerable environmental plasticity of the taxa. In addition, there are several levels of polyploidy, and within these, strains may behave consistently differently within a single species. Much confusion has centred around R. penicillatus: this species is now thought to have arisen through hybridisation. However, with careful attention to detail, most collections may be determined.

Taxonomic characters

i) Leaves. The presence or absence of Laminate leaves is an important taxonomic character. R. hederaceus and R. omiophyllus develop laminate leaves only, whereas R. fluitans, R. penicillatus var. calcareus, R. circinatus, and R. trichophyllus never have laminate leaves. Beware of forms of R. aquatilis growing on mud or in shallow ditches. These often develop no floating leaves, and have often been misidentified as R. trichophyllus.

   Laminate leaf shape is rather variable, and young leaves are similar in shape. Lobing is generally constant, but mature or senescent leaves may have more lobes than is normal for the species. Whilst the character of the submerged divided leaf is often of importance, terrestrial divided leaves have no diagnostic value.

ii) Flowers. The blue tips to the sepals of R. baudotii is a diagnostic character. Sepal size, sepal petal-length ratio, and whether or not sepals are reflexed are all useful features.
   The shape of nectar pits is sometimes a useful character, but only large, healthy petals should be examined.
   The receptable remains globose in all British species, except R. baudotii, in which it elongates. It is usually densely hairy in all species except R. fluitans, R. hederaceus and R. omiophyllus in which it is glabrous or only sparsely hairy. A receptacle with more than 10 bristle-like hairs is regarded as hairy. Frequently even glabrous receptacles develop soft, downy hairs. These hairs should be disregarded unless in excess of about 25.

iii) Peduncle. The length of the peduncle is a good taxonomic character. Some variation occurs when plants are deeply submerged, but generally the peduncle petiole ratio remains constant in a particular species.

1. Laminate leaves present.

2. Plant with no finely-dissected submerged leaves.

3. Leaves not deeply-divided (i.e. to ½ or less); mature achenes not winged; sepals without blue tips; receptacle not elongating at maturity.

4. Leaf-lobes broadest at the base; sepals and petals approximately equal in length; sepals not reflexed. R. hederaceus L.

4. Leaf-lobes narrowest at the base; petals 2-3 times as long as the sepals; sepals reflexed. R. omiophyllus Ten. (R. lenormandii F.W. Schultz)

3. Leaves usually deeply three-lobed (i.e. to 2/3 or more); mature achenes winged; sepals almost always with blue tips; receptacle elongating at maturity. R. baudotii Godr.

2. Plant with finely-dissected submerged leaves.

5. Submerged leaves yellowish-green; laminate leaves usually deeply three-lobed; immature achene glabrous; mature achene winged; receptacle elongating at maturity. R. baudotii

5. Submerged leaves not yellowish-green; laminate leaves usually five-lobed; immature achene pubescent; mature achene not winged; receptacle not elongating at maturity.

6. Fruiting peduncle shorter than the petiole of the opposed laminate leaf; petals less than 10mm; nectar pits circular. R. aquatilis L.

6. Fruiting peduncle longer than the petiole of the opposed laminate leaf; petals more than 10mm; nectar pits pyriform.

7. Mature submerged leaves shorter than the internodes; leaves not more than 10cm, the segments usually rigid and divergent, more than 6 divisions to each leaf. R. peltatus Schrank.

7. Mature submerged leaves longer than the internodes; leaves to 30cm long, the segments flaccid, sub-parallel, often less than 6 divisions to each leaf. R. penicillatus (Dum.) Bab. var. penicillatus (R. peltatus subsp. pseudofluitans of CTW)

1. Laminate leaves absent. (reliable when at least three flowers present).

8. Leaf segments held rigidly in one plane. R. circinatus Sibth.

8. Leaf segments not held rigidly in one plane.

9. Leaves yellowish-green; mature achenes winged; sepals almost always with blue tips. R. baudotii

9. Leaves not yellowish-green; mature achenes not winged; sepals rarely with blue tips.

10. Petals less than 5mm, not contiguous at anthesis; nectar-pits lunate. R. trichophyllus Chaix

10. Petals more than 5mm, contiguous at anthesis; nectar-pits pyriform or circular.

11. Mature leaves as long or longer than internodes, the segments subparallel; nectar-pits pyriform.

  12. Receptacle densely pubescent; leaves to 25cm long, the segments rounded. R. penicillatus var. calcareus (R.W. Butcher) Cook

  12. Receptacle glabrous; leaves to 50cm long, the segments obconical. R. fluitans Lam.

11. Mature leaves shorter than the internodes, the segments spreading; nectar-pits pyriform or circular.

  13. Sepals less than 5mm; petals less than 10mm; fruiting peduncle less than 50mm; nectar-pits circular. R. aquatilis L.

  13. Sepals more than 5mm; petals more than 10mm; fruiting peduncle often more than 50mm; nectar-pits pyriform. R. penicillatus var. calcareus

based on Nature Conservancy Report, N.T.H. Holmes, 1979.

Ranunculus fluitans occurs mainly in central Britain, and is much rarer than R. penicillatus in the region. R. penicillatus var. penicillatus occurs mainly in western mainland Britain (e.g. Cumbria, Wales, S.W.England) and is the dominant in Irish rivers. R. penicillatus var. calcareus is more widespread and occurs in less flashy rivers, and in meso- to eutrophic waters.

PAPAVER L.

Two closely related species are P. dubium and P. lecoqii. The capsule character given in CTW is not reliable.

Latex white; anthers violet; seeds bluish-black. P. dubium L.

Latex turning yellowish on exposure to air; anthers yellow; seeds chocolate-brown. Mainly on calcareous soil. P. lecoqii Lamotte

A.B. Mowat & S.M. Walters, in Flora Europaea Vol.1 (1964)

FUMARIA L.

Determination of Fumaria species is sometimes difficult. Poorly-grown, or shaded specimens often produce whitish, cleistogamous flowers much smaller than is normal, and normally recurved pedicels may be straight. Sepal shape, however, is fairly constant, even in shaded or poorly-grown plants. In well-grown plants, other useful characters are raceme length, peduncle length, and the number of flowers per raceme. Characters in CTW found not to be very helpful are:

1) The erect margin of lower petals of F. muralis subsp. boraei. This is often distinctly spreading.
2) The lateral compression of the upper petal.
3) Apical pits of fruits.
4) Leaf characters, except for a few of the small-flowered species.

1. Sepals 4-6.5 x 2-3mm, oval or oblong, obtuse to acute.

2. Corolla 10-12(14)mm, creamy-white, wings and tips blackish-red. Pedicels strongly reflexed in fruit. Sepals usually ± oval.
F. capreolata L.

2. Corolla 10-13mm, pinkish-purple, wings and tips dark-purple. Pedicels not strongly reflexed in fruit. Sepals usually ± oblong.
F. purpurea Pugsl.

1. Sepals 2-5 x 1-3mm.

3. Corolla 9-12mm.

4. Sepals broadly-ovate, 3-5 x 2-3mm, acute or acuminate. Corolla magenta-pink, with dark tip to upper petal.
F. muralis Sond. ex Koch subsp. boraei (Jord.) Pugsl.

4. Sepals ovate, 3 x 1mm, acute, with forwardly-directed teeth ± all round the margin. Corolla salmon-pink, rarely with dark tip to upper petal.
F. bastardii Bor.

3. Corolla 7-8mm.

5. Robust. Flowers usually more than 20 per raceme. Sepals 2.5-3.5 x 1-1.5mm. Fruit distinctly flat-topped to emarginate.
F. officinalis L. subsp. officinalis

5. Less robust. Flowers usually less than 20 per raceme. Sepals about 2 x 1mm. Fruit faintly apiculate.
F. officinalis L. subsp.wirtgenii (Koch)Arcangeli

With regard to specific differences between F. bastardii and F. muralis subsp. boraei, in addition to sepal-size, flower colour is usually a good guide. The corolla of F. muralis is usually magenta-pink with a dark tip to the upper petal, whilst F. bastardii is usually more salmon-pink, and only rarely has a dark tip to the upper petal. Fruit shape is a good character once one is familiar with the difference. The fruits of F. bastardii are slightly broader than those of F. muralis subsp. boraei. The relative width of the base in proportion to the width of the fruit is similar in both, but what is important is the shape of the seed near the base; there is a more distinct 'neck' to the base of the F. muralis subsp. boraei fruit.

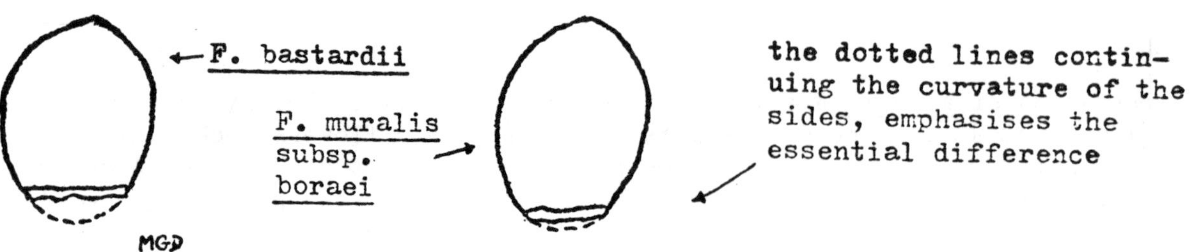


F. officinalis is the most frequent species in the eastern part of the region, whilst F. muralis subsp. boraei appears to be the commoner of the two in the west. Other species are generally scarce.

M.G. Daker, pers. comm. (1979)

Page 15 Sepals of Fumaria species x5

Sepals taken from several specimens, photocopied, and enlarged.

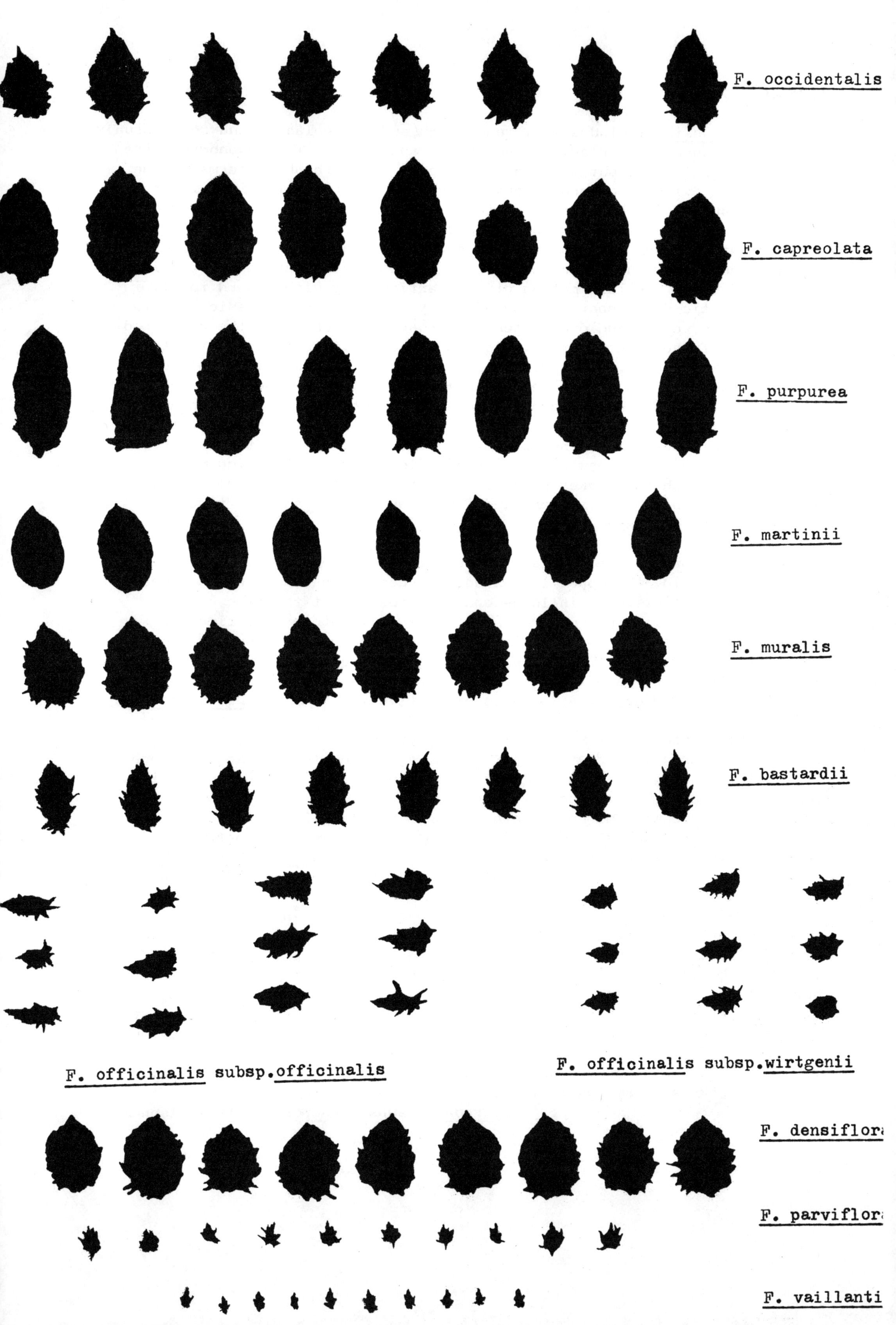
F. occidentalis
F. capreolata
F. purpurea
F. martinii
F. muralis
F. bastardii
F. officinalis subsp. officinalis
F. officinalis subsp. wirtgenii
F. densiflora
F. parviflora
F. vaillanti

BRASSICA L.

The genus Brassica includes many vegetables and crop-plants, and some casuals, the latter found particularly around ports and docks. The taxonomy of the genus is complicated, but need not trouble us if we think in terms of the Cabbage, the Swede, the Turnip, and their wild counterparts, ignoring the scores of subspecies and varieties into which these species have been divided. Difficulties have arisen in separating them, largely because most keys fail to take into account the considerable changes which affect a) the radical leaves, and b) the inflorescences of all three species during the growing season. Some of the information given, particularly with regard to a), is only diagnostic for a short period of the year, i.e. many Brassicas begin growth in one season and overwinter as a rosette which forms the early radical leaves of the plant in the following season. It is these leaves to which most keys refer. In spring, these rosettes die away long before the time of flowering and cannot be used unless the plant has been examined throughout the whole of the growing season. Sometimes (as in B. rapa) a summer rosette or partial rosette is formed, and is quite different from that which disappeared in the spring. In all three species the inflorescence lengthens to some extent during the flowering and fruiting period. The rate at which this takes place is important, and can best be seen by noting the proportion of flowers and fruits present at any one time on a particular inflorescence.

Leaves bluish-green, the radical leaves with undulate margins but hardly lyrate-pinnatifid, the upper leaves not widened at the base and only one third clasping the stem; inflorescence lengthening at the time of flowering into a long raceme, reaching its full length by the time the flowers fade, i.e., at any one time the lengthened raceme consists almost entirely of flowers, or is at the same stage of fruit formation throughout; petals large, pale lemon-yellow, twice as long as the sepals; all stamens erect. In our area, a relic of cultivation. B. oleracea L.

Leaves all glaucous, the radical leaves lyrate-pinnatifid, disappearing before the plant flowers, the upper leaves expanded at the base, and more than half clasping the stem; flowers remain until the corymb expands into a short raceme, i.e., buds overtop the flowers, and mature fruits are present at the same time on the stem below; petals yellow or pale-orange, nearly twice as long as the sepals; outer stamens shorter than inner, curved outwards at the base. B. oleracea L. x B. rapa L. (B. napus L., incl. var. napobrassica (L.) Rchb.)

Leaves glaucous, the radical leaves in a rosette lyrate-pinnatifid, grass-green and bristly, but often disappearing very early and succeeded by an almost glabrous set, upper leaves with a broadened, deeply-cordate base more than half clasping the stem; flowers more nearly in a corymb, those newly-opened overtopping the buds; the corymb lengthens into a short raceme after the flowers have fallen, i.e., the lengthened raceme consists mainly of fruits; petals bright-yellow, much shorter than in the previous two species, and about $1\frac{1}{2}$ times as long as the sepals; outer stamens about 1/2 length of inner, curved outwards at the base. Widespread, especially on river-banks and damp waste land.
B. rapa L. subsp. campestris (L.) Clapham

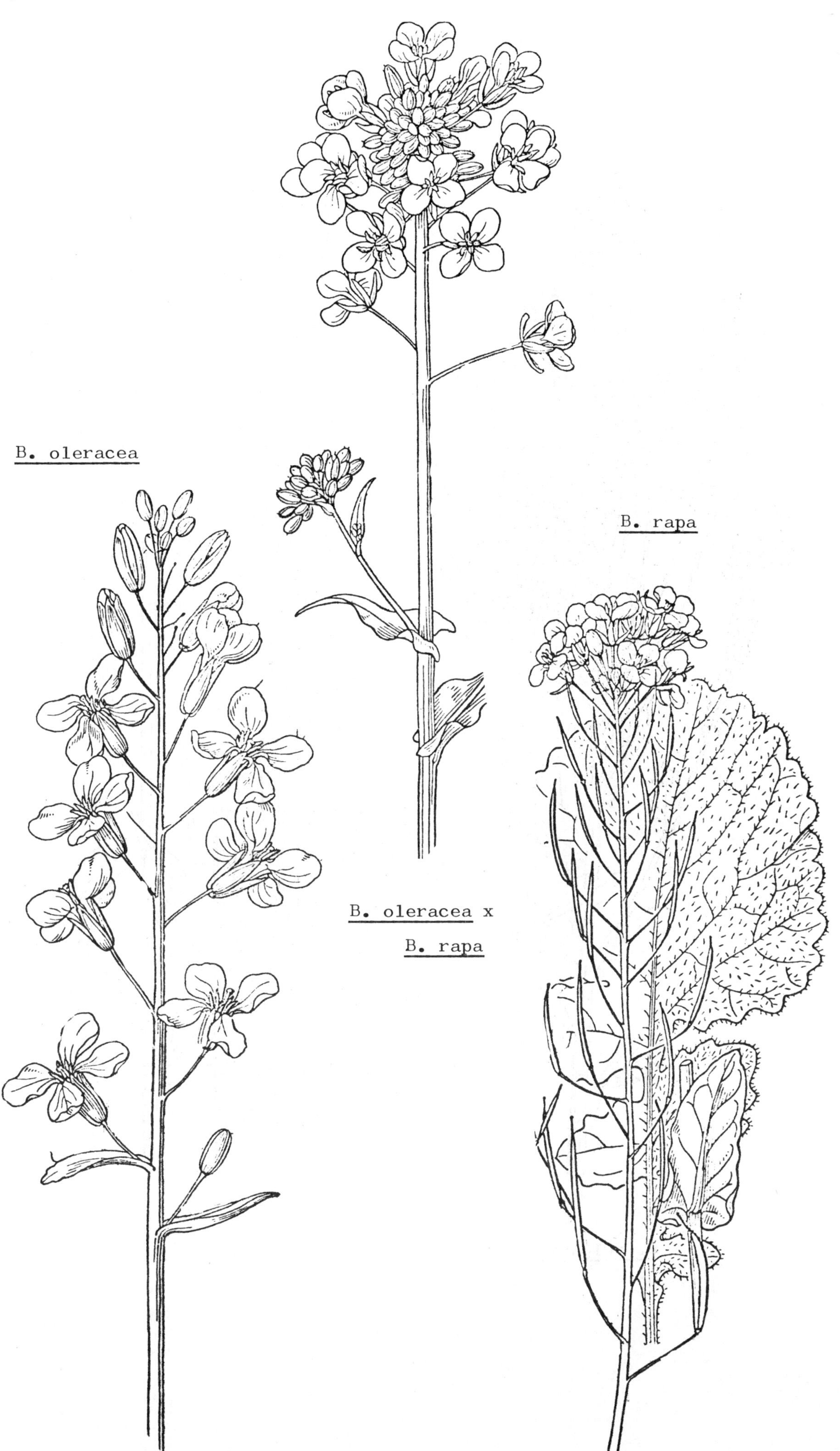
B. oleracea
B. rapa
B. oleracea x
B. rapa

B. vulgaris

B. stricta

B. intermedia

BARBAREA R.Br.

Four species of Barbarea have been recorded in the region. B. vulgaris is common and widespread, the remaining three species rare.

1. Upper stem-leaves simple, toothed, or shallowly-lobed.

2. Upper stem-leaves obovate, usually with 1-few narrow, lateral lobes near base; flower buds glabrous; petals about 1.5 x as long as sepals; mature siliquae 15-25mm long. B. vulgaris R.Br.

2. Upper stem-leaves ovate, coarsely sinuate-lobed; flower buds at least sparsely hairy; petals about 2 x as long as sepals; mature siliquae 20-30mm. B. stricta Andrz.

1. Upper stem-leaves pinnately-lobed.

3. Petals to 2 x as long as sepals; mature siliquae 10-30mm. B. intermedia Bor.

3. Petals about 3 x as long as sepals; mature siliquae 30-60mm. B. verna (Mill.) Aschers.

The basal leaves are too variable to be of use in separating B. intermedia and B. verna. Length of petals and siliquae afford much the best characters.

## SINAPIS L.

*Sinapis alba* is scarce in the region, though perhaps sometimes overlooked as *S. arvensis*. The two are readily separated, however.

Upper leaves sessile, usually simple, lanceolate, coarsely-toothed: beak of siliqua conical, ± straight, rather more than half as long as the valves. *S. arvensis* L.

All leaves petiolate, lyrate-pinnatifid; beak of the siliqua strongly-compressed, sabre-shaped, equalling or exceeding the valves. *S. alba* L.

## RAPHANUS L.

Although these two species can normally be separated readily on habit, leaf-shape, petal-colour, etc., yellow-flowered plants of *R. raphanistrum* in coastal areas can be very similar to small plants of *R. maritimus*. It is, therefore, best to distinguish these on the consistently distinct fruiting characters.

Siliqua 3-6mm. in diameter, with distinct, but not very deep, constrictions between the 3-8 seeds, breaking readily into 1-seeded units; beak slender, up to 5 times the length of the uppermost unit. Plant 20-60cm. Petals white, lilac, or yellow, with darker veins. *R. raphanistrum* L.

Siliqua 5-8mm. in diameter, with short, deep constrictions between the 1-5 seeds, not readily breaking into 1-seeded units; beak slender, rarely more than 2 times as long as the uppermost unit. Plant 20-80 cm. Petals usually yellow. *R. maritimus*

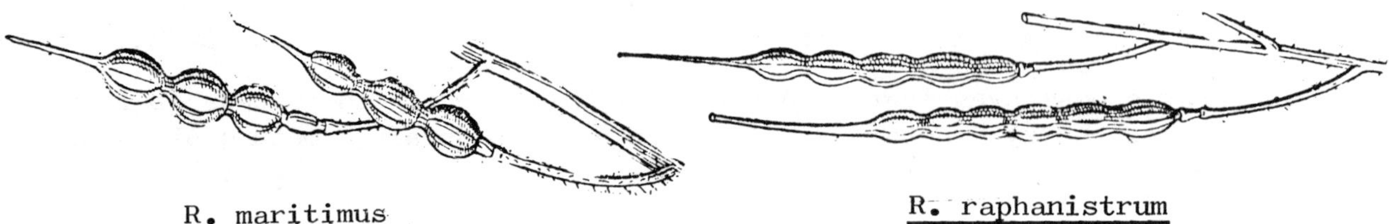

*R. maritimus* *R. raphanistrum*

LEPIDIUM L.

Of the six species recorded in our area, only L. heterophyllum is widespread. The shapes of basal and cauline leaves have been used as key characters, but most specimens have lost the basal leaves by the time they set fruit; also the leaf shape is rather variable,as is the hairiness of the plant. More reliable characters are the size and shape of the mature siliculae (from the middle of the raceme), but even here the range often much exceeds that given in floras.

1. Silicula 4mm or more, with a prominent wing connate with the lower part of the style; upper leaves amplexicaul.

2. Annual or biennial; silicula densely covered with small vesicles; style usually not projecting from notch. L. campestre (L.) R.Br.

2. Perennial; silicula with few or no vesicles; style projecting from notch. L. heterophyllum Benth. (L. smithii Hook.)

1. Silicula usually less than 4mm, with style free from the narrow wing (except sometimes in L. sativum) or wing absent; upper leaves usually not amplexicaul.

3. Leaves strongly dimorphic, the upper ovate and amplexicaul (appearing perfoliate), the lower bipinnate; petals yellow, longer than the sepals. Casual. L. perfoliatum L.

3. Leaves not strongly dimorphic, although the upper sometimes lanceolate and gradually passing into the lower bipinnate ones; petals white or absent.

4. Petals present, exceeding the sepals; silicula 2-5 x 2-6mm, with or without an apical notch; style projecting or not; leaves more than 2mm wide.

5. Silicula 2mm, orbicular, unwinged, with a very small apical notch, or none; style projecting; plant 50-130cm, inflorescence a large, dense, pyramidal panicle. L. latifolium L.

5. Silicula 3-5 x 5-6mm, broadly elliptic to orbicular, narrowly-winged above, with a large apical notch; style not projecting; plant less than 50cm; inflorescence not a large, dense panicle. L. sativum L.

4. Petals usually absent; silicula 2.0-2.5 x 1.5-1.8(2.0)mm; mid- and upper leaves linear, entire, less than 2mm wide; racemes forming dense, intricate, rather compact inflorescence. Casual. L.ruderale L.

T.B. Ryves, Watsonia 11(4) 367-372 (1977)

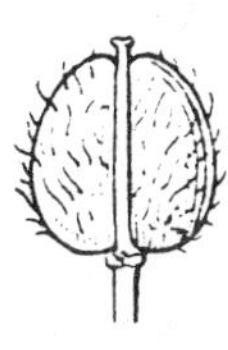

L.latifolium

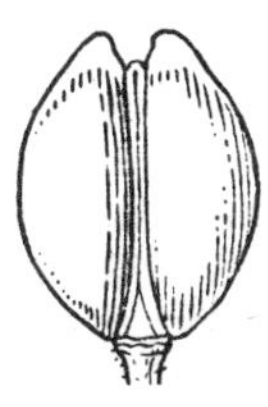

L.ruderale

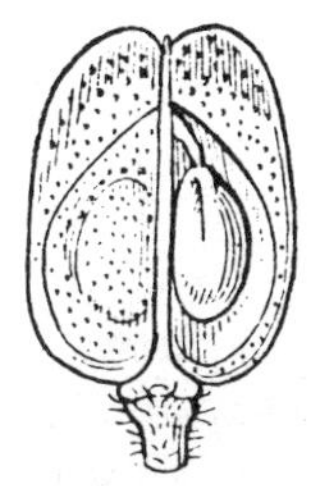

L.campestre

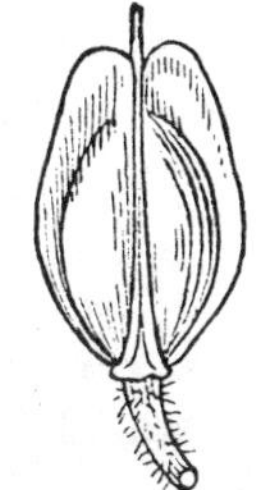

L.heterophyllum

Fruit - all x8

COCHLEARIA L.

The status of the inland populations of *Cochlearia* is much disputed, and have hitherto been recognised as belonging to *C. alpina* (Bab.) Wats., *C. pyrenaica* DC., *C. officinalis* L., or *C. micacea* Marshall. The characters formerly used to separate these taxa were mostly quantitative, or plastic, or both.

Recent cytotaxonomic studies of Gill et al, Watsonia 12(1), 15-21 (1978), have shown that one diploid and two tetraploid cytotypes exist. The diploid (2n = 12) occurs at moderate altitudes and appears restricted to base-rich habitats, and thus corresponds ecologically and cytologically with *C. pyrenaica*, and British material named as such by Gill et al. However, no characters have yet been found to distinguish it morphologically from tetraploid *C. officinalis* and is, therefore, best included in that species. The two tetraploids are distinguishable morphologically and cytologically. The 2n = 24 cytotype corresponds with *C. officinalis*, and the 2n = 26 with *C. micacea*, the latter not found in our area. Plants hitherto named *C. alpina* appear to be indistinguishable morphologically and cytologically from *C. officinalis* sensu stricto, and should perhaps be regarded merely as inland ecotypes of *C. officinalis*.

*C. officinalis*, 2n = 12, inland in Durham and Westmorland

2n = 24, coastal and inland sites, widespread

EROPHILA VERNA (L.) Chevall.

The two British segregates are separated as follows;

Sepals 1.5-2mm; petals 2.5mm; stamens longer, exceeding stigma; silicula 6-10mm; elliptical to oblanceolate; fruiting pedicels 10-25mm.
*E. verna* subsp. *verna*

Sepals 1mm; petals 2mm; stamens shorter than stigma; silicula not more than 5mm; ovate to suborbicular; fruiting pedicels 3-18mm.
*E. verna* subsp. *spathulata* (Láng) Walters

S.M. Walters, in *Flora Europaea*, Vol.1, (1964)

CARDAMINE L.

All three of the British small-flowered species occur in the region.

*C. impatiens* is readily distinguished by its cauline leaves having basal, stipule-like auricles clasping the stem (sagittate base). It is rare in the area, and only a few localities are known in Westmorland and Cumberland, mainly woodland sites on limestone.

The remaining two species are best separated on the number of stamens, and less reliably on habitat preference. *C. flexuosa* has 6 stamens (but two of them may be small and inconspicuous), and is mainly a plant of moist and shady places. *C. hirsuta* has 4 stamens, and shows a preference for the drier, more open sites.

In older plants, a flexuous stem with numerous stem leaves will usually distinguish *C. flexuosa* from *C. hirsuta* which has a ± straight stem with very few stem leaves.

RORIPPA Scop.

R. islandica in Britain consists of two species. R. islandica (Oeder ex Murray) Borbás subsp. islandica is known from a few scattered localities in Scotland, Ireland, and the Isle of Man. R. palustris (L.) Besser subsp. palustris is widespread in Britain, though infrequent in the region.

1. Siliqua 9-18mm. Petals twice as long as the sepals. R. sylvestris (L.) Besser

1. Siliqua less than 9mm. Petals shorter than, or longer than the sepals.

2. Petals about twice as long as the sepals. Silicula 3-6 x 1-3mm, ovoid; pedicels 16-18mm. Robust plant, with a stout, erect stem. R. amphibia (L.) Besser

2. Petals equalling, or shorter than the sepals. Siliqua 4-9 x 1.5-2mm, ± oblong, pedicels 4-10mm long. Plant usually fairly small.

3. Siliqua often large, with pronounced, rectangular valves, 2-3 times as long as their pedicels; sepals less than 1.6mm; seeds very faintly colliculate (having rounded swellings). R. islandica subsp. islandica

3. Siliqua not more than 1.2 times as long as their pedicels; sepals more than 1.6mm; seeds more coarsely colliculate. R.palustris subsp.palustris

B. Jonsell, Symb.bot.Upsal., 19(2),1-22 (1968)

RORIPPA Scop.

Note that the following two species are separated on the character of the seed coat, and that the number of rows of seeds is not a reliable character.

R.nasturtium-aquaticum (L.) Hayek (Nasturtium officinale R.Br.)
Fruit relatively short (13-18mm.), broad and straight. Dehiscence of the long stamens introrse (facing inwards). Seeds with about 25 depressions on the surface. Leaves and stems remaining green in autumn.

R. microphylla (Boenn.) Hyland (Nasturtium microphyllum (Boenn.) Reichb.)
Fruit longer (16-22mm.), narrow and usually curved. Dehiscence of the long stamens extrorse (facing outwards). Seeds with about 100 depressions on the surface. Leaves and stems turning purple-brown in autumn.

R. x sterilis Airy Shaw (R.microphylla x R. nasturtium-aquaticum)
Fruit dwarfed and deformed, though sometimes with 1 or 2 seeds intermediate in seed-coat between those of the parents. Leaves and stems turning purple-brown in autumn. Stamens in all directions, often deformed and twisted. Pollen mostly abortive.

Fruit x20

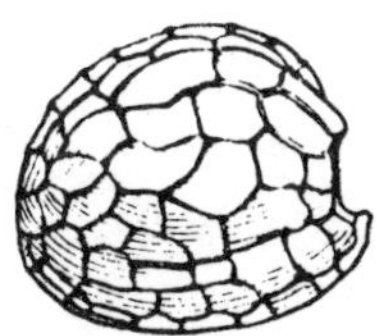

R.nasturtium-aquaticum

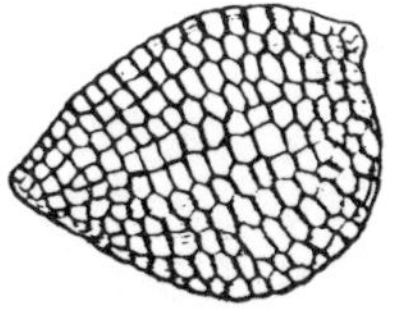

R.microphyllum

VIOLA L.

Species of Viola are, in the main, not difficult to identify, but careful examination of plants is often necessary.

V. odorata and V. hirta can usually be separated on a combination of leaf-shape, length of pedicel, flower-scent, and corolla-colour, but all these are variable. Both species have hairy leaves and pedicels. However, the presence of stolons in V. odorata, and their absence in V. hirta is diagnostic, and this character can be used when the plant is either purely vegetative or in flower.

V. rupestris is known at only two sites in the region, and is nationally rare. It is distinct from the other hairy violets in its caulescent habit (at least some leaves cauline), the ovate-orbicular leaves, and the pubescent capsule. The pubescence of peduncle and petiole is characteristically fine. It is, however, often difficult to separate from the hybrid V. riviniana x V. rupestris (V. x burnatii) which occurs with the species in Durham. The hybrid is generally intermediate between the parents in leaf- and flower characters, and the pubescence is even finer than that in the true V. rupestris. No seed is set.

V. riviniana and V. reichenbachiana are sometimes confused, and they cannot be separated vegetatively. The former species has sepals becoming accrescent in fruit with large appendages, and corolla with a pale-lilac, stout spur. The latter species has sepals (obsolete in fruit) with very small appendages, and a dark-lilac, slender spur. The hybrid V. riviniana x V. reichenbachiana has been recorded in Durham, and is doubtless much overlooked in the region. It is generally intermediate between the parents, but unless compared with them in the field, it is often difficult to identify. Its high infertility, however, is a useful character. Few good pollen-grains are produced, and very little seed set.

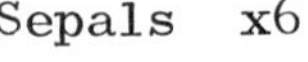
Sepals x6

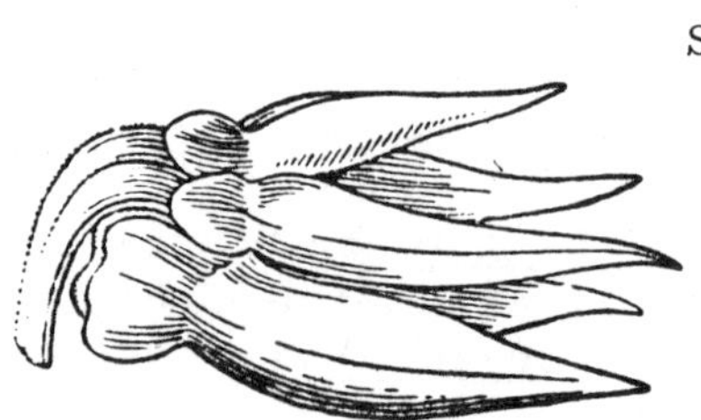
V. riviniana

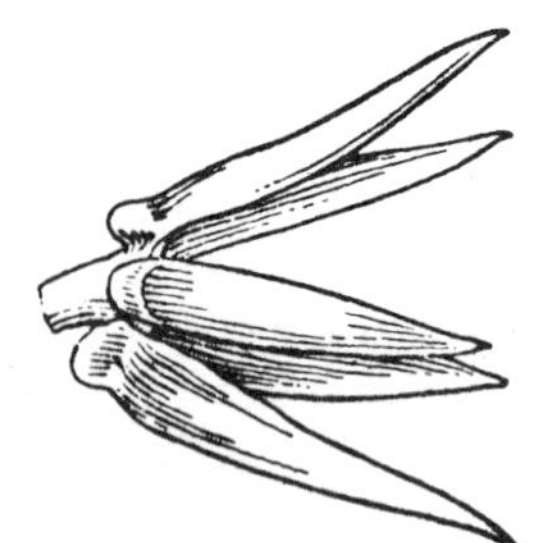
V. reichenbachiana

V. canina has leafy flowering stems, but no basal rosette. No other species occurring in the region has this combination of characters. The corolla is blue, and has a yellowish spur. It is fairly frequent in Cumbria, occurring inland and on the coast. Elsewhere it is more or less confined to the coast, and is very local even there.

Key to plants with floral axis present, but no open flowers.

1. Plant with stipules lanceolate to ovate-lanceolate, not leaf-like.

2. Plant acaulous (i.e. with leaves and pedicels all radical).

3. Plant with leafy, procumbent, rooting stolons; leaves and petioles hairy. V. odorata L.

3. Plant without stolons; leaves and petioles glabrous or hairy.

4. Plant with long, creeping rhizomes; leaves glabrous, orbicular-reniform; petiole sometimes hairy. V. palustris L.

4. Plant without rhizomes; leaves hairy, triangular-ovate to oblong-ovate; petiole hairy. V. hirta L.

2. Plant caulescent (i.e. with leafy flowering stems).

5. Plant with a basal rosette of leaves.

6. Leaves, petioles, and peduncles with fine pubescence. (small plants may appear acaulous). V. rupestris Schmidt

6. Whole plant glabrous.
V. riviniana Rchb., and V. reichenbachiana Jord. ex Bor.

5. Plant without a basal rosette of leaves; all leaves cauline, ovate to ovate-lanceolate, cordate at base. V. canina L.
(also V. lactea and V. stagnina, but not in region)

1. Plant with stipules leaf-like: pinnatifid, palmatifid, or at least deeply-lobed.

7. Plant with long, slender, creeping rhizomes, producing solitary, unbranched flowering stems. V. lutea Huds.

7. Plant without rhizomes (or very short and not creeping), producing usually much-branched flowering stems.

8. Mid-lobe of stipule usually lanceolate, ± entire, not leaf-like.
V. tricolor L.

8. Mid-lobe of stipule usually ovate-lanceolate or ovate, crenate-serrate, leaf-like. V. arvensis Murr.

Stipules V. tricolor V. arvensis

## ELATINE HEXANDRA/MONTIA FONTANA

These two species are superficially similar in gross morphology, and both are plants of wet places - Elatine submerged, or on wet mud, and Montia mainly in flushes, wet pastures, streamsides, etc. Both species have inconspicuous flowers, but are very distinct in flowering and fruiting characters. Elatine hexandra is usually rather smaller than Montia fontana, and is much rarer, though several sites are known in vc 69 and 70.

| | Elatine hexandra (Lapierre) DC. | Montia fontana L. |
|---|---|---|
| habit | prostrate or decumbent, 2.5-10cm. | decumbent to ± erect. |
| stem | rooting at the nodes. | rooting at the nodes in aquatic forms. |
| stem leaves | opposite, spathulate, 2-8cm long. | opposite, spathulate to obovate, 7-20mm long. |
| stipules | present | absent |
| flowers | axillary, solitary in the leaf axils. | terminal, though may appear lateral through overtopping by a branch. |
| sepals | 3 | 2 |
| petals | 3, free. | 5, unequal, joined in a short tube which is cleft to the base. |
| stamens | 6 | usually 3, opposite the 3 smaller petals and adhering to their bases. |
| seeds | 8-12 in each cell, curved, elongate | 3, flattened-ovoid |
| flowering | July to September | May to October |

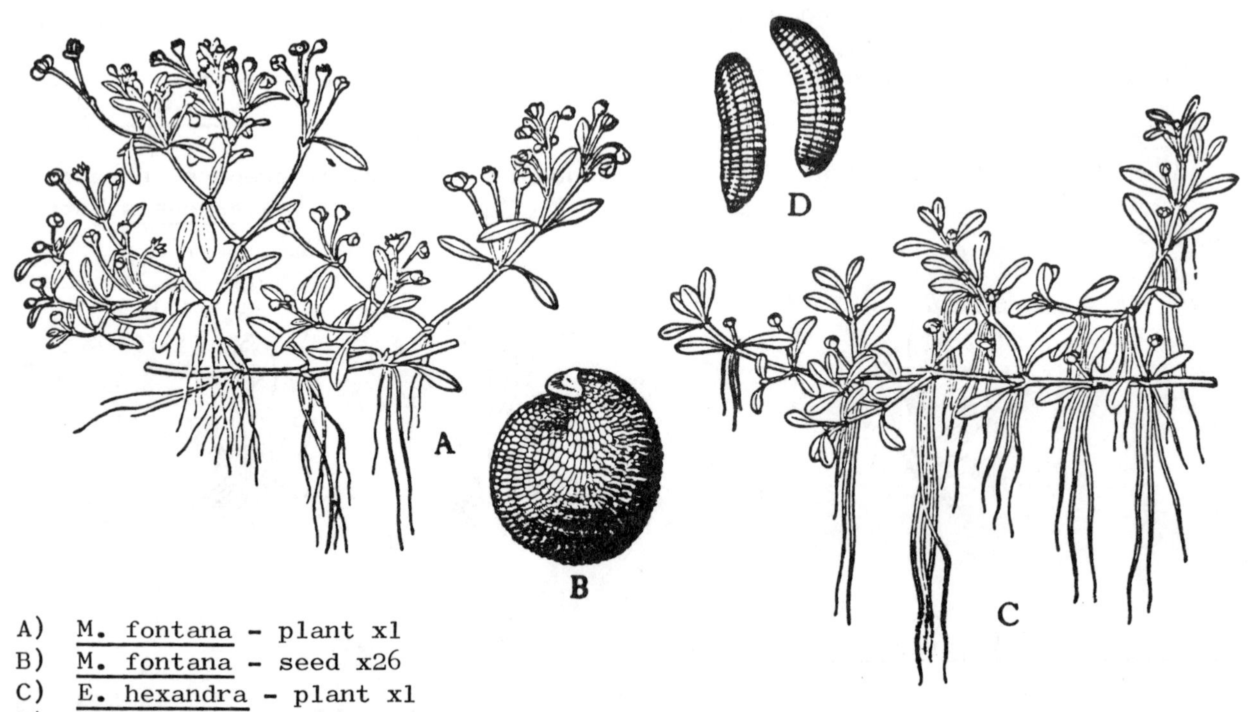


A) M. fontana - plant x1
B) M. fontana - seed x26
C) E. hexandra - plant x1
D) E. hexandra - seed x30

HYPERICUM x DESETANGSII Lamotte
(H. perforatum L. x H. maculatum Crantz)

This hybrid between H. perforatum and H. maculatum is likely to occur where the parent species grow in close proximity. The characters of H. x desetangsii are rather vague because its parents appear to be interfertile, and back-cross readily, thus producing a ± continuous range of forms between the parental species.

The description of the hybrid in CTW is highly misleading, and the main characters are given here.

i) The second pair of stem lines in H. maculatum usually disappear in the hybrid, but they may be evident.

ii) The leaves usually show intermediate characters between the densely reticulate venation with few or usually no pellucid glands of H. maculatum, and the lax reticulation with numerous glands of H. perforatum.

iii) The sepals are the most useful indicators. The intermediate condition is medium-width sepals with a few superficial dark glands, the apex is obtuse-erose; although the whole range between the parents exist, the hybrids apparently always have some denticules.

N.K.B. Robson, pers. comm, 1974.

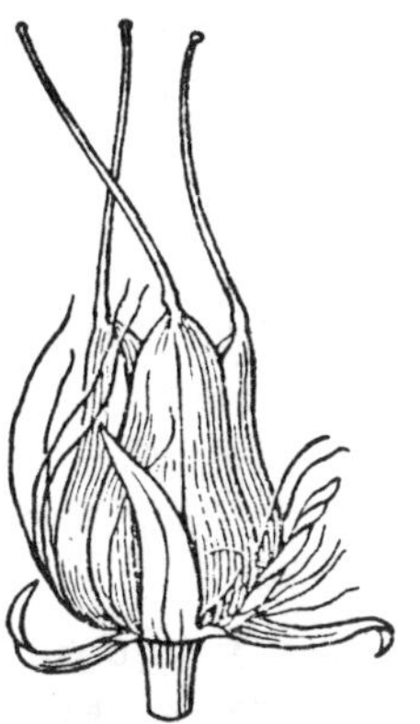

H. perforatum

H. maculatum

SILENE x INTERMEDIUM (Schur) Druce
(S. alba (Mill.) E.H.L. Kranse x S. dioica (L.) Clairv.)
(Melandrium album x M. rubrum)

This hybrid, with pink flowers and intermediate in other characters, is often seen. It is moderately fertile with 0-30% inviable pollen; this fertility combined with backcrosses to both parents can produce a continuous spectrum of intermediates.

SILENE ALBA (Mill.) E.H.L. Krause/S. NOCTIFLORA L.
(Melandrium album/M. noctiflorum)

These two species are best separated on floral characters. S. alba is dioecious (i.e. each flower has either styles (5), or stamens present, but not both in the same flower). S. noctiflora is monoecious (i.e. each flower has both styles (3) and stamens present).

Both species may be evening-scented, and have flowers rolled inwards during the day, though these are features mainly of S. noctiflora.

S. noctiflora is glandular-viscid, especially above, but sometimes not markedly so. It is local in the region, occurring in lowland arable fields.

## CERASTIUM TOMENTOSUM L.

Plants named C. tomentosum or C. biebersteinii occur as garden escapes in Britain, and both may occur in our area. They have been much confused, not least because most of their characters overlap. Both are white-lanate, and very variable in habit, hairiness, and leaf-shape.

Capsule teeth patent, with revolute margins; leaves 10-30 x 2-5mm ; sepals 5-7mm long. — C. tomentosum L.

Capsule teeth ± erect, with flat margins; leaves 20-50 x 3-8mm ; sepals 6-10mm long. — C. biebersteinii DC.

## CERASTIUM DIFFUSUM Pers. / C. SEMIDECANDRUM L.

Care is needed in the determination of this pair of species, especially with depauperate specimens.

| | C. diffusum (C. atrovirens Bab.) (C. tetrandrum Curt.) | C. semidecandrum L. |
|---|---|---|
| bracts | entirely herbaceous | at least the upper bracts with broad, scarious margins |
| flowers | usually 4-merous, sometimes 5-merous | 5-merous |
| petals | about 1/5 bifid, 3/4 as long as sepals | slightly notched, 2/3 as long as sepals |
| stamens | 4(5) | 5 |
| styles | 4(5) | 5 |
| pedicels | much longer than sepals | usually equalling or slightly exceeding sepals |
| fruiting pedicels | usually erect throughout | at first sharply deflexed from the base |

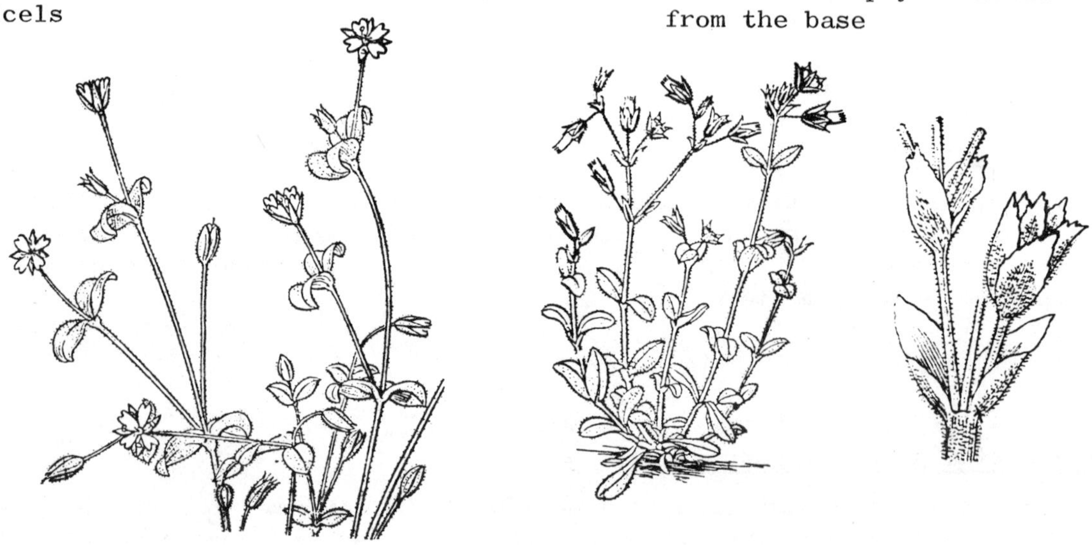

C. diffusum is mainly a coastal plant in our region, but is sometimes found inland on railway ballast. C. semidecandrum is also mainly coastal, though also occurs inland on calcareous and sandy soils, and on railway ballast.

STELLARIA NEGLECTA Weihe/S. MEDIA (L.) Vill.

Woodland and shade forms of S. media are often very large and can be mistaken for S. neglecta. The two are readily separated, but S. neglecta, though certainly scarce in our region, may sometimes be overlooked.

Stamens usually 10 ; pedicels long, slender, ± straight, usually deflexed in fruit ; seeds (1.1)1.3-1.6mm. S. neglecta Weihe

Stamens 3-8 : pedicels downwardly-curved in fruit, often flexuous, not long and slender; seeds 0.8-1.4mm. S. media (L.)Vill.

Characters of little use are sepal-size (large overlap), and seed colour.

STELLARIA PALLIDA (Dum.) Piré/S. MEDIA (L.) Vill.

This species can be confused with small and slender, apetalous forms of S. media (L.) Vill. and can normally be recognised only in the spring. (March to May).
The number of stamens is much the best character - 90% of the plants have 2 stamens. Much less reliable is the size and colour of the seeds.
It has been much overlooked, and may be locally frequent in coastal regions, where small, dense, yellow green mats with plants spreading from the centre should be examined.

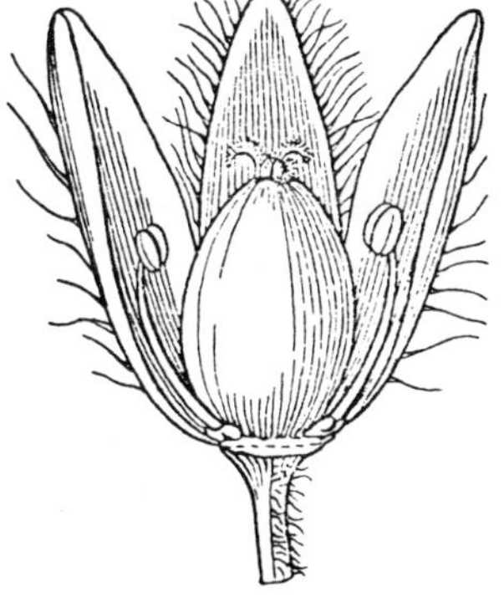

SAGINA L.

Because of the small size of Sagina species, careful attention to detail is required.

1. Mat-forming perennial with short, non-flowering main stem bearing a central leaf rosette; lateral stems ascending from procumbent, rooting bases. S. procumbens L.

1. Annual, without vegetative stems at time of flowering. Main stem flowering, branches not rooting, decumbent to erect.

  2. Leaves awned. Mean seed size less than 0.4mm.

    3. Fruiting sepals appressed to slightly patent, subacute; terminal capsule of well-grown plants about 1.25 times as long as the sepals. Mean seed size more than 0.34mm. S. apetala Ard. subsp.apetala (S. ciliata Fries)

    3. Fruiting sepals patent, subobtuse; terminal capsule of well-grown plants more than 1.25 times as long as the sepals; mean seed size less than 0.34mm. S. apetala Ard. subsp.erecta (Hornem.)Hermann. (S. apetala auct.)

  2. Leaves obtuse to apiculate, but not awned. Mean seed size more than 0.4mm. S. maritima Don.

A.R. Clapham of N. Jardine, in Flora Europaea Vol.1 (1964)

SAGINA NODOSA (L.) Fenzl/MINUARTIA VERNA (L.) Hiern

A superficial glance can lead to mis-identification of this pair of species.

S. nodosa has petals nearly 2x as long as the sepals; styles 5; leaf clusters in the axils of the upper leaves give the stem a knotted appearance. Generally in the damper places on sand, peat etc.

M. verna has petals little longer than the sepals; styles 3; anthers red; stems lacking dense axillary clusters of leaves. Plant of the drier places, on base-rich rocks, screes, and open pastures, and particularly characteristic of old lead spoil heaps.

ARENARIA SERPYLLIFOLIA L.

Two subspecies of the complex occur in the region. Subsp. leptoclados is scarce throughout, though probably under-recorded. It is usually more delicate in all its parts.

Capsules distinctly swollen at base, exceeding 3mm long, fracturing when pressed at maturity; pedicels rather stout, about 0.5mm thick, straight; sepals ovate-lanceolate. A. serpyllifolia L. subsp. serpyllifolia

Capsules ± straight-sided, less than 3mm long, possible to indent without fracturing when mature; pedicels slender, about 0.3mm thick, upturned near the tip; sepals lanceolate.
A. serpyllifolia subsp. leptoclados (Reichenb.) Guss.

subsp. serpyllifolia

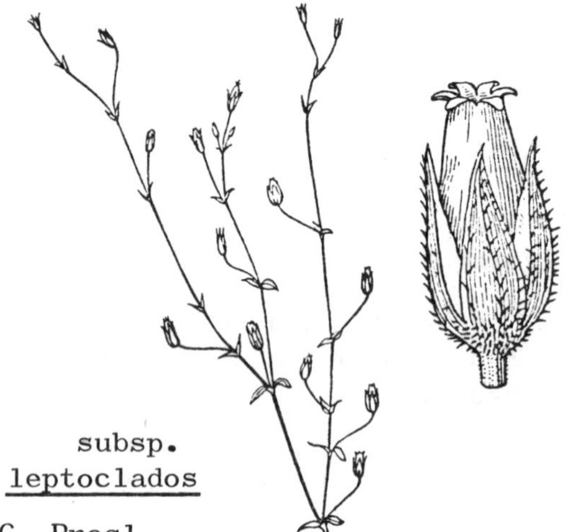

subsp. leptoclados

SPERGULARIA MARINA (L.) Griseb./S.MEDIA (L.) C. Presl.

These two species can be very similar. The seed wing is often too variable to be of decisive diagnostic value in their separation, and some populations of S. media have completely unwinged seeds. Fruit and calyx lengths vary widely too, and are of limited use.

Perennial, with a thick, woody stock; sepals usually 4-6mm long; petals equalling or shortly exceeding sepals; stamens 10, rarely 7-9.
S. media (L.) C. Presl.
(S. marginata Kittel)

Annual, with a ± slender tap-root; sepals usually 2.5-4mm long; petals not exceeding sepals; stamens 1-5(8). S. marina (L.) Griseb.
(S. salina J. & C. Presl.)

P. Monnier & J.A. Ratter, in Flora Europaea Vol.1 (1964)

MONTIA FONTANA L.

All four subspecies are found in our region, and identification requires microscopic examination of the seed. For this reason, they are all greatly under-recorded, and little is yet known of their distribution and ecology.

Plants with seeds (generally found from May to October) should be collected, dried in an envelope, and sent to a recorder who has the facilities for seed-examination.

1. Ripe seeds 1.1-1.35mm, smooth, shining. subsp. fontana

1. Ripe seeds 0.85-1.2mm, dull or somewhat shining, with at least some low tubercles on keel.

2. Ripe seeds dull, entirely covered with broad, obtuse tubercles; usually annual, with erect, caespitose stems. On damp, sandy ground. Ripe seeds 1.0-1.2mm. subsp. chondrosperma (Fenzl) S.M.Walters

2. Ripe seeds somewhat shining; tubercles confined to keel; annual or perennial, of varied habit.

3. Ripe seeds 0.85-1.1mm, with 3-4 rows of long, acute tubercles on keel. subsp. amporitana Senner (subsp. intermedia (Beeby) S.M.Walters)

3. Ripe seeds 0.9-1.1mm, with variably developed broad, low tubercles on keel. subsp. variabilis S.M.Walters

S.M.Walters, in Flora Europaea Vol.1 (1964)

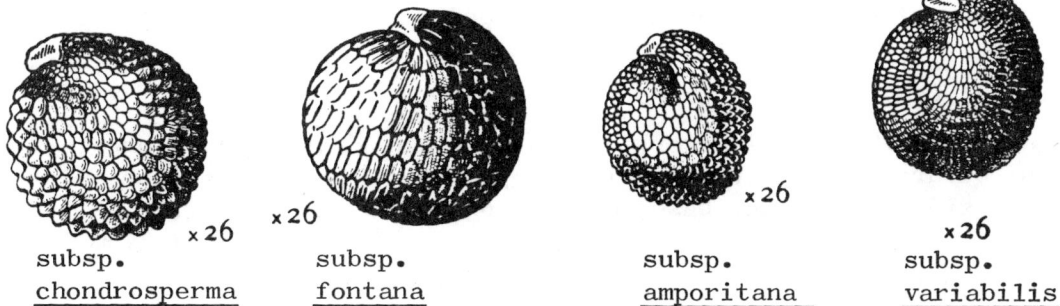

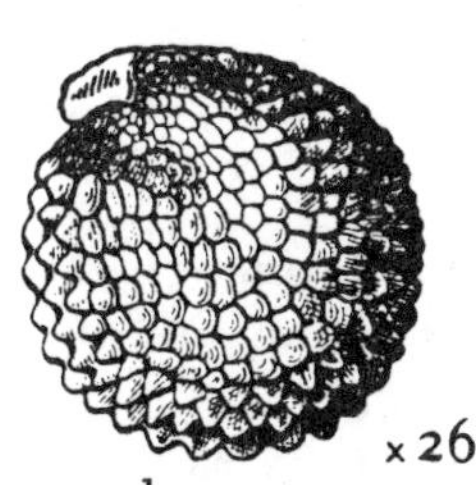

subsp. chondrosperma

subsp. fontana

subsp. amporitana

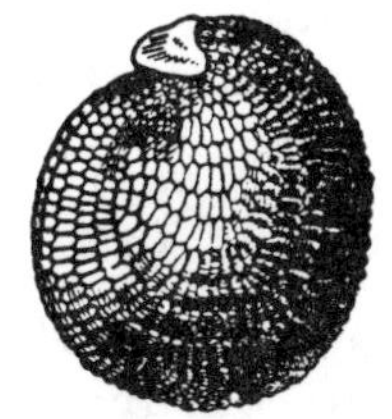

subsp. variabilis

Seeds

CHENOPODIUM L.

The identification of Chenopodium species has always presented recorders with problems, mainly because of the great phenotypic plasticity of many species, and the necessity of examining the seeds microscopically (x40, at least). Most species, therefore, are safely determined only late in the season, when ripe seed is present.

The marking of the seed testa is taxonomically important. To see this, the skin-like pericarp which closely invests the seed must be removed. This may be done simply by rubbing the seed ("pericarp easily removable"), or in some species only by boiling the seed, or scraping with a needle ("pericarp persistent"). This persistence, or non-persistence may be seen in the two widespread species - C. bonus-henricus and C. album respectively.

1. Perennial, with triangular-hastate, or -sagittate leaves; inflorescence narrowly-pyramidal and tapering; stigmas 0.8-1.5mm, exserted; seeds vertical except in terminal flowers; perianth persistent. C. bonus-henricus L.

1. Annual, leaves variable, but not hastate or sagittate; inflorescence not narrowly pyramidal, stigmas not more than 0.5mm; seeds vertical or horizontal; perianth persistent or not.

2. Inflorescence axis and perianth glabrous; seeds horizontal or vertical.

3. Leaves entire, or rarely with a single inconspicuous tooth on each side, ovate-elliptic, cuneate at base; stems 4-angled; seeds horizontal, black. C. polyspermum L.

3. Leaves, except the uppermost, not entire (very rarely entire in C. rubrum, but then seeds red-brown); stems ridged but not 4-angled; seeds vertical, red-brown.

4. Leaves densely glaucous-farinose beneath, green above, coarsely sinuate-dentate to sinuate. C. glaucum L.

4. Leaves green or reddish, glabrous beneath, usually coarsely serrate-lobed. C. rubrum L.

2. Inflorescence axis and perianth ± conspicuously farinose, at least when young; seeds usually all horizontal.

5. Plant smelling strongly of decaying fish; leaves entire, or with a single angle on one or both sides towards the base. C. vulvaria L.

5. Plant not smelling of decaying fish; leaves not dentate or lobed.

6. Pericarp very persistent; seeds dull, with a sharp, rather prominent keel; testa with minute, very close, rounded pits; leaves often strongly-dentate, not at all 3-lobed. C. murale L

6. Pericarp easily removed; seeds with obtuse margins; leaves variable; testa not with minute, rounded pits.

7. Seeds 1.0-1.15mm in diameter; testa with close, radially-elongate pits; leaves usually with a single, prominent, divergent lobe on either side towards the base, middle lobe ± parallel-sided. C. ficifolium Sm.

7. Seeds 1.2-1.6(1.85)mm in diameter; testa faintly radially striate, otherwise nearly smooth; leaves very variable. C. album L.

Sculpturing of seed testa

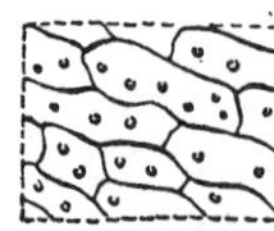
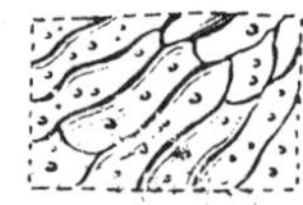

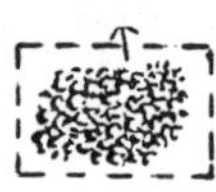
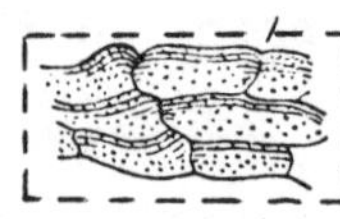
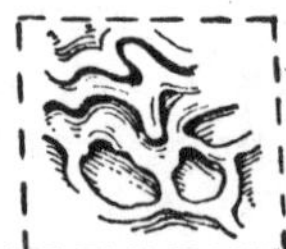

a. C.polyspermum
b. C.vulvaria
c. C.album
d. C.ficifolium
e. C.murale
f. C.rubrum
g. C.bonus-henricus

All species, except C.bonus-henricus and C.album, are rare, and many of the identifications are suspect. Immature Atriplex patula may be mistaken for a Chenopodium.

ATRIPLEX L.

Like Chenopodium, mature fruit is needed for the identification of most species of Atriplex, and this is usually not present until late summer or autumn. The British representatives of this genus have been much misunderstood, especially the pair, A. prostrata (A. hastata) and A. glabriuscula, because of inadequate or erroneous descriptions in texts and papers.

A. praecox and A. longipes have recently been discovered in Britain, and the latter is known in Westmorland.

1. Leaf venation appearing (at x12) as a conspicuous dark-green reticulate pattern when the leaf surface is scraped with a razor-blade. (Fig.a) Bracteoles tough-cartilaginous. Whole plant white or silvery. A. laciniata L.

1. Leaf venation (at x12) not showing a dark, reticulate pattern when the leaf surface is scraped with a razor-blade. (Fig.b.) Bracteoles not cartilaginous. Plant neither white nor silvery.

2. Lower leaves linear or lanceolate.

3. Lower leaves linear, without basal lobes; bracteoles ovate, thick, margins united only at the base, apices lingulate and reflexed at maturity. Coastal halophyte. A. littoralis L.

3. Lower leaves lanceolate with up-curving basal lobes; bracteoles rhombic, thin, margins united almost to the middle, apices acute to acuminate, not reflexed at maturity. Non-halophytic, inland and coastal weed of disturbed ground. A. patula L.

2. Lower leaves triangular or rhombic-ovate.

4. Lower leaves rhombic-ovate; mature plants small (mostly 8-10cm high). Restricted to the lower littoral zone of coastal beaches. Not found in our region. A. praecox Hülphers

4. Lower leaves triangular; mature plants larger (mostly more than 20cm high). Occurring throughout the littoral zone of coastal beaches, and inland.

5. Bracteoles rhombic, margins united up to the middle, spongy-thick from the middle to the base; seed radicle strongly up-pointing (Figs i, j). Coastal halophyte. A. glabriuscula Edmonston

5. Bracteoles ovate or triangular, margins united only at the base, thin or evenly-thickened; seed radicle out-pointing, or obliquely up-pointing (Figs k-m). Coastal and inland.

6. Lower leaves deltoid-triangular, less than twice as long as wide, base truncate to sub-cordate (basal angle more than 160°); bracteoles thin or evenly-thickened, sessile; axillary bracteoles rare, similar to terminal ones. Coastal and onland. A. prostrata Boucher ex DC. (A. hastata L.)

6. Lower leaves narrowly-triangular, about twice as long as wide base cuneate (basal angle less than 145°); axillary bracteoles common, morphologically different from terminal ones, thin, with pedicels 10-20mm. Restricted to coastal and estuarine salt-marshes usually in tall vegetation. A. longipes Drejer

Atriplex longipes x A. prostrata is recognised by Taschereau in Durham, and hybrids between A. prostrata and A. glabriuscula may be frequent where both parents occur in close proximity.

/drawings overleaf

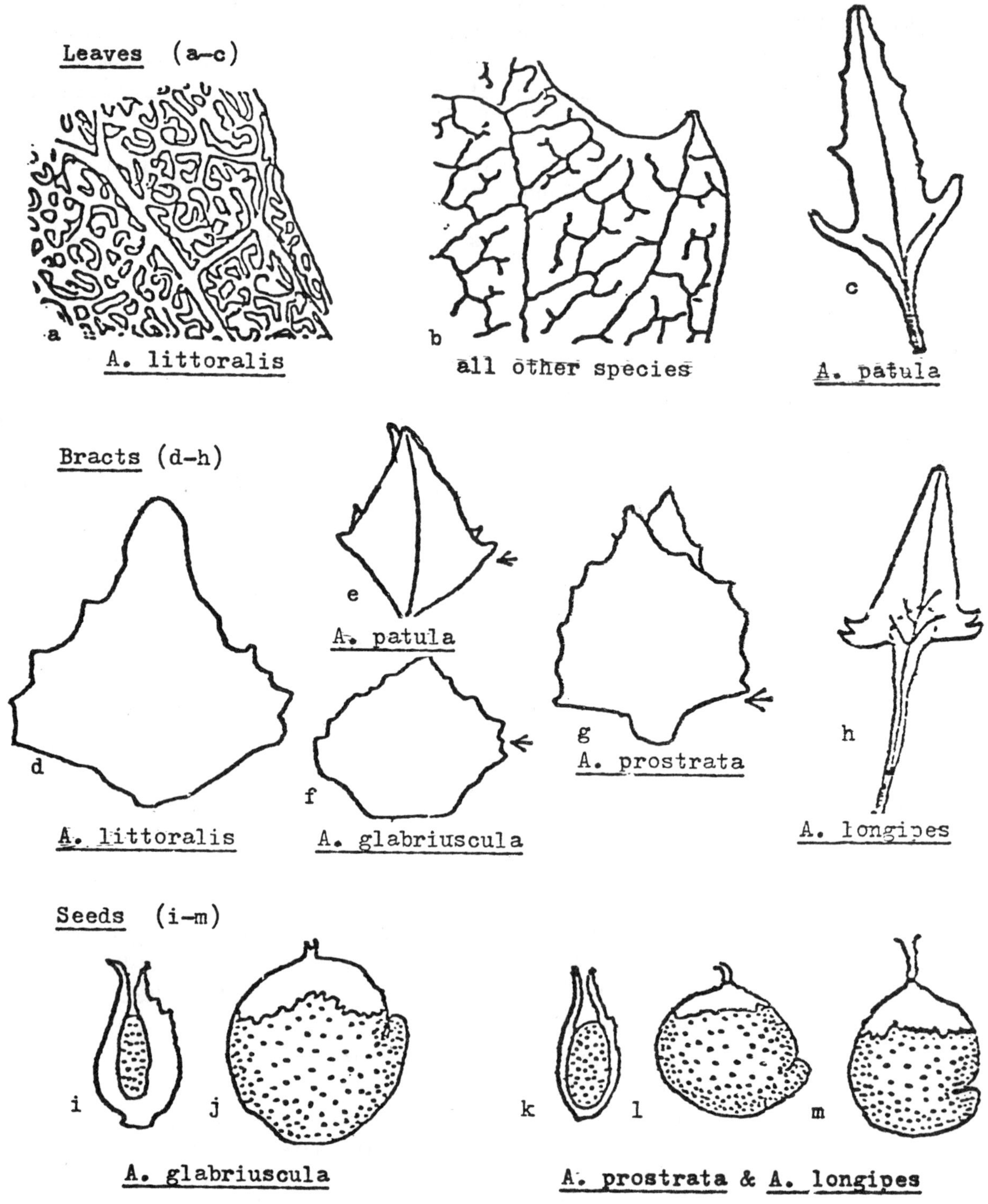


In e, f, and g, the margins of the bracteoles are united from the base to the arrow-mark.

i and k show the bracteole and seed in section.

j, l, and m, show the seed in front view.

P.M. Taschereau, pers. comm. (1978)

SALICORNIA L.

Only four species of this poorly-understood genus have been found in our region. Phenotypic variation is very great, and the habit of individual plants may be greatly changed by many factors, among which are:

i) the nature of the substrate or rooting medium;
ii) degree of crowding and competition;
iii) shading;
iv) damage or loss of root support.

Identification should only be attempted on well-grown plants with an undamaged main stem in September and October, and several plants should be selected from a population (single plants are of little use). Typical coloration does not develop fully until the seed is ripe. It is usually impossible to identify dried material.

Flora Europaea places S. perenne in Arthrocnemum (A. perenne (Miller) Moss), all other British species remaining in Salicornia. The genus Salicornia is divided into two main groups, the characters of which are tabulated below.

Characters of species with 3-flowered cymes

| | europaea group | procumbens group |
|---|---|---|
| colour in fruit | usually becoming red or purple | becoming pale-green, yellow, or brownish |
| uppermost primary branches | straight, making an angle of (usually) less than 45° with the main stem | curving upwards |
| terminal spike | with 3-12(22) fertile segments | with (4)6-30 fertile segments |
| fertile segments | with convex sides | usually ± cylindrical, or slightly concave sides |
| lateral flowers | appearing much smaller than the central | almost equalling the central |
| flowers | often cleistogamous | chasmogamous |
| stamens | usually 1, rarely 0 or 2 | 1-2 |
| anthers | 0.2-0.5mm, usually dehiscing before exsertion, or not exserted | (0.5)0.6-1mm, dehiscing after exsertion |
| seeds | 1-1.7mm | 1.4-2.2mm |

Key to British species

1. Perennial, uprooted with difficulty, with prostrate, woody stems, and ascending, fleshy branches; with many non-flowering branches. Arthrocnemum perenne (Mill.) Moss (S. perennis Mill.)

1. Annual, uprooted with ease; all branches terminated by an inflorescence. (Salicornia species)

2. Cymes 1-flowered; infructescence disarticulating. S. pusilla J. Woods

2. Cymes 3-flowered; infructescence not disarticulating. (europaea group, and procumbens group).

europaea group

1. Lower fertile segments of the terminal spike 3-5mm wide at the narrowest point; upper edge with an inconspicuous scarious margin, not more than 0.1mm wide; plant glaucous- or grass-green (sometimes becoming yellow-green), usually with red colour appearing diffusely in the fertile segments. S. europaea L.

1. Lower fertile segments of terminal spike up to 3.5(4)mm wide at the narrowest point; upper edge with a 'conspicuous' scarious margin about 0.1-0.2mm wide; plant dark-green (sometimes becoming yellow-green, with dark purplish-red first appearing around the flowers and along upper edge of fertile segments, sometimes eventually colouring the whole segment. S. ramosissima J. Woods

procumbens group

1. Lower fertile segments of the terminal spike at least 2-3.5mm wide; plant usually becoming light brownish- or orange-purple. Not recorded in the region. S. nitens P.W. Ball & Tutin

1. Lower fertile segments of the terminal spike usually exceeding 3.5mm wide; plant usually without purple coloration (rarely pale red or purple).

2. Terminal spikes with 6-15(22) fertile segments, ± cylindrical; spikes of the primary lateral branches cylindrical; plant dull green to yellowish-green, often becoming bright-yellow in fruit. Found in v.c. 69. S. fragilis P.W. Ball & Tutin

2. Terminal spikes with 12-30 fertile segments, distinctly tapering towards the apex; spikes of the primary lateral branches tapering; plant dark, dull green becoming paler or dull-yellow in fruit. S. dolichostachya Moss

(S. lutescens is now considered a variant of S. fragilis , and S. obscura a variety of S. europaea.

P.W. Ball, in Flora Europaea Vol. 1 (1964)

TILIA L.

Care needs to be taken in distinguishing T. cordata from T. x vulgaris (T. x europaea auct.), particularly in woodland in Cumbria where both species occur. T. x vulgaris is also commonly planted in parks, gardens, and on roadsides throughout the region.

Cymes obliquely erect; tertiary veins of leaves not prominent; hair-tufts in vein-axils reddish-brown; fruit about 6mm, ribs obscure or absent. Leaves cordate at base. T. cordata Mill.

Cymes pendent; tertiary veins of leaves prominent; hair-tufts in vein-axils whitish; fruit about 8mm, slightly ribbed. Leaves generally larger, less cordate at base. T. x vulgaris Hayne

GERANIUM MOLLE L./G. PUSILLUM L./G. ROTUNDIFOLIUM L.

These three species can look very similar; they can be separated on a combination of floral and fruiting characters. Leaf shape is very variable.

G. rotundifolium is a casual in the region. G. pusillum is scarce throughout, but is perhaps overlooked.

1. Petals entire; sepals ± patent during flowering; mericarps (carpels) hairy; seeds pitted. G. rotundifolium L.

1. Petals emarginate; sepals ± erect during flowering; mericarps glabrous or hairy; seeds smooth.

 2. Mericarps (excluding style) glabrous; petals 3-7mm; all stamens with anthers. G. molle L.

 2. Mericarps (excluding style) hairy; petals 2-4mm; 3-5 of the stamens without enthers i.e. reduced to staminodes. G. pusillum L.

D.A. Webb & I.K. Ferguson, in Flora Europaea Vol.2 (1968)

ERODIUM L'Hérit.

Of the species in this genus, only E. cicutarium subsp. cicutarium is widespread, though local and mainly coastal. There are old and/or unlocalised records of E. maritimum, and this species should be looked for in coastal areas. Subsp. dunense of E. cicutarium is no longer regarded as distinct.

1. Leaves simple, lobed. Small, decumbent annual. E.maritimum (L.)L'Hérit.

1. Leaves pinnate. Plants usually larger.

 2. Apical pits of mericarps(carpels) eglandular, furrowed or not at base; leaflets pinnatifid, pinnatisect, or 2-pinnatifid.

  3. Mericarps 4-5mm, without or with a very faint furrow below the apical pit; flowers usually not more than 7mm in diameter, pale pink, petals without a dark basal ptach; plant usually densely glandular-pubescent. E. cicutarium (L.) L'Hérit. subsp. bipinnatum Tourlet (E. glutinosum Dum.)

  3. Mericarps 5-7mm, with a distinct furrow below the apical pit; flowers usually 12mm or more in diameter, purplish-pink, petals often with a dark basal patch; plant eglandular or glandular, pubescent. E. cicutarium (L.)L'Hérit.subsp.cicutarium

 2. Apical pits of mericarp glandular, with a deep furrow at the base: plant glandular, usually robust, but sometimes dwarf; most of the leaflets divided to less than half-way. E. moschatum (L.) L'Hérit.

D.A. Webb & A.O. Chater, in Flora Europaea Vol.2 (1968)

Mericarps of E. cicutarium and E. moschatum

OXALIS L.

Six species have been recorded in our region, all but one (O. acetosella) are naturalised and established aliens, and specimens of all these aliens should be collected for identification. The whole plant, including the root system should be collected.

1. Petals yellow.

2. Stem rooting at nodes; leaves alternate; stipules auriculate.

3. Tap-root not thick. Leaves green; leaflets 3-5 x 3-6mm; flowers solitary; fertile stamens 5, the others reduced to staminodes; capsule 5-8mm , about 3x as long as wide, abruptly narrowed to the tip. O. exilis A. Cunn.

3. Tap-root thick, about 2-5mm in diameter. Leaves usually suffused purple; leaflets 5-18 x 8-23mm; inflorescence 1-8 flowered; fertile stamens usually 10; capsule 12-20mm , at least 5 times as long as wide, tapering to the tip. O. corniculata L.

2. Stem not rooting at the nodes; leaves mostly sub-opposite; stipules not auriculate.

4. Inflorescence umbellate; fruiting pedicels deflexed; stipules oblong; stem with non-septate hairs. O. stricta L.

4. Inflorescence not umbellate; fruiting pedicels not deflexed; stipules absent; stem with septate hairs. O. europaea Jord.

1. Petals white, lilac or purple.

5. Stem rhizomatous; bulb absent, though rhizome may be swollen; petals white. O. acetosella L.

5. Stem not rhizomatous, erect, bearing sessile, axillary bulbils; bulb present; petals lilac with darker veins. O. incarnata L.

D.P. Young, in Flora Europaea Vol.2 (1968)

ULEX L.

All three native species of Ulex occur in our region. U.gallii is much more frequent in the west, and U. minor is found only in Cumberland, at a few sites near Carlisle. The lengths of the calyx and petals provide the most satisfactory characters by which U. gallii and U. minor may be separated. Several calices and/or petals should be measured to obtain mean lengths. The key in CTW is inadequate.

1. Calyx densely hairy, with spreading hairs; bracteoles much wider than pedicels; primary spines often bearing branches (and flowers) a considerable distance from their bases; main flowering period in spring. U. europaeus L.

1. Calyx less densely hairy, with appressed hairs; bracteoles not, or only slightly wider than pedicels; main flowering period in autumn.

2. Calyx 6-9.5(11)mm long; standard petal (5)7.5-11.5(12.5)mm long. U. minor Roth.

2. Calyx (9)10-13(14)mm long; standard petal (10.5)12-16.5(18)mm long. U. gallii Planch.

M.C.F. Proctor, Watsonia 6(3)177-187 (1965)

ULEX x DOUIEAE Druce (U. europaeus L. x U. gallii Planch.)

Hybrid bushes are intermediate between the parents, the most useful characters being the pedicel-thickness, the size of the bracteoles and pod, the number of ovules (7-10), the indumentum of the calyx, and the flowering time (August to March, maximum in September).

The hybrid appears to be generally highly fertile, and extensive hybrid swarms occur locally. It is known in S.W. England and in N. Wales, and may well occur in the west of our area. Samples should be collected of plants which appear intermediate.

M.C.F. Proctor, in Stace (1975)

ONONIS L.

Plant rhizomatous and usually procumbent; stems rooting above the base, hairy all round; leaflets obtuse or emarginate. Flowers pink, wings ± equalling keel. Pods shorter than the calyx, which enlarges in fruit. Widespread in grassland and dunes. O. repens L.

Plant not rhizomatous, usually erect or ascending; stems not rooting above the base, with two lines of hairs (examine a cross-section); leaflets obtuse or acute. Flowers brighter pink, wings shorter than keel. Pod exceeding calyx. Local, preferring clays and heavy soils. O. spinosa L.

The presence or absence of spines is not a diagnostic character and must not be used alone to distinguish O. repens and O. spinosa. O. repens is often spiny (var. horrida Lange) and O. spinosa is sometimes without spines (var. mitis Gmel.).

According to J.K. Morton, many intermediate populations exist along the Durham coast, presumably O. repens x O. spinosa.

MELILOTUS OFFICINALIS (L.) Pall./M. ALTISSIMA Thuill.

Both M. altissima and M. officinalis are local in our area, and are differentiated on a combination of floral and fruiting characters.

Ovary and young legume glabrous; legume 3-5mm, with transverse veins, brown when ripe; wings and standard equal, longer than the keel. M. officinalis (L.) Pall.

Ovary and young legume pubescent; legume 5-6mm, reticulate-veined, black when ripe; wings, standard, and keel equal. M. altissima Thuill.

A. Hansen, in Flora Europaea Vol.2 (1968)

TRIFOLIUM DUBIUM Sibth./T. MICRANTHUM Viv.

Depauperate plants of T. dubium may be mis-identified as T. micranthum, which is normally the smaller plant. The most reliable characters appear to be those of the pedicel.

Pedicels rather stout, shorter than the calyx-tube; leaves to 11mm, terminal leaflet petiolulate; heads 3-15(25) flowered; corolla 3-3.5mm.
T. dubium Sibth.

Pedicels slender, as long or longer than the upper limb of the calyx-tube; leaves to 5(8)mm, terminal leaflet subsessile; heads 1-6 flowered; corolla 2-3(4)mm.
T. micranthum Viv.

D.E. Coombe in Flora Europaea does not differentiate between the two species on the basis of notching or otherwise of the standard petal (see CTW) and it would seem unwise to use this distinction.

D.E. Coombe, in Flora Europaea Vol.2 (1968)

T. dubium

T. micranthum

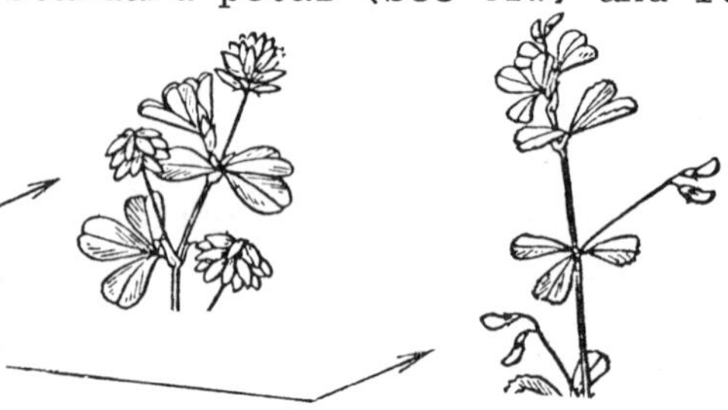

ANTHYLLIS VULNERARIA L.

A very polymorphic species, divisible into several infra-specific taxa (many of which are frequently recognised as species), between which many intermediates occur. The taxa, (treated as subspecies here), are generally poorly-differentiated. Subsp. vulneraria is by far the commonest taxon in the region.

1. Calyx 2-4(5)mm wide, the lateral teeth small, appressed to the upper teeth (often visible only when fresh); bract-lobes narrowly-deltate, acute; leaflets of upper cauline leaves equal.

2. Indumentum of stems composed entirely of patent hairs. Leaves confined to lower part of stem, fleshy, glabrous above; calyx concolorous.
subsp. corbierei (Salmon & Travis) Cullen

2. Indumentum of stems with at least some appressed hairs, calyx usually with a red apex. By far the commonest taxon in the region.
subsp. vulneraria

1. Calyx (4.5)5-7mm wide, the lateral teeth obvious, not appressed to the upper; bract-lobes parallel-sided, obtuse; leaflets of upper cauline leaves unequal.

3. Calyx with sparse, appressed, silky hairs. Waste places, etc.
subsp. carpatica (Pant.)Nyman

3. Calyx with spreading, shaggy hairs. Upland areas, especially on limestone.
subsp. lapponica (Hyl.) Jalas

The calyx measurements refer to the flowers at anthesis (first opening), and measurements made later than this are misleading as the calyx expands considerably in fruit. The character of the calyx teeth is difficult until one's eye is in.

J. Cullen, in Flora Europaea Vol.2. (1968)

LOTUS L.

L. corniculatus is very variable; tall, ascending forms can be mistaken for L. uliginosus. L. uliginosus is generally taller, with longer leaflets and petioles, and peduncles, though there is a considerable overlap.

Stem solid, or nearly so; calyx teeth erect in bud, the upper pair separated by an obtuse sinus. Pastures and grassy places. L. corniculatus L.

Stem hollow; calyx teeth spreading in bud, the upper pair separated by an acute sinus. Marshes and damp grassy places.
L. uliginosus Schkuhr (L.pedunculatus auct.)

VICIA SATIVA L. agg.

The taxonomy of this species aggregate has been much confused in the past, owing in part to a large number of cultivated strains having escaped and intermingled with the native populations. Quantitative characters mostly vary in such a way that the four British segregates, V. lathyroides, V. angustifolia, V. segetalis, and V. sativa, differ by increasing size of most of their parts (and in that order). However, there is a wide measure of overlap so that these characters are not alone sufficient for distinguishing the species. The shape of the leaflets on the lower leaves of the flowering shoot is often markedly different from those on the upper leaves which bear flowers in their axils. Plants with this character well-developed are heterophyllous.

Main differentiating characters:

1. Seeds tuberculate. V. lathyroides L.
   Seeds smooth.

2. Pods constricted between seeds. V. sativa L.
   Pods not constricted between seeds.

3. Plant markedly heterophyllous. V. angustifolia L.
   Plant not, or scarcely heterophyllous. V. segetalis Thuill.

Diagnoses

V. lathyroides L. Strongly heterophyllous; tendrils simple; flowers 6-9mm , concolorous (usually dull purple); pods 18-30mm., brown to black, smooth, glabrous; seeds tuberculate.

V. angustifolia L. Strongly heterophyllous; tendrils branched; flowers 14-19mm , concolorous (usually bright pink); pods 23-28mm, brown to black, smooth, glabrous; seeds smooth.

V. segetalis Thuill. More or less isophyllous; tendrils branched; flowers 9-26mm, bicolorous (with standard petal paler than the wings); pods 28-70mm, brown to black, smooth, usually glabrous; seeds smooth.

V. sativa L. More or less isophyllous; tendrils branched; flowers 11-26mm, bicolorous (with standard petal paler than the wings); pods 36-70mm, yellowish to brown, constricted between the seeds, often pilose; seeds smooth.

E. Hollings and C.A. Stace, Watsonia 12(1) 1-14 (1978)

## RUBUS CAESIUS L.

This species has procumbent, very pruinose, terete weak stems, ternate leaves and lanceolate stipules; prickles usually few, weak. Prickles short and aciculate on the upper branches. Drupelets as well as stems covered with a whitish bloom (pruinose); drupelets few in number, 2-5(10) and loosely adherent to the receptacle. This species, although fairly common in the magnesian limestone denes and woodland, and on carboniferous limestone has sometimes been recorded in error for very similar members of Section *Triviales* of the *R. fruticosus* agg. Brambles in Sect. *Triviales* are suspected hybrids of *R. caesius* with members of the *R. fruticosus* agg., and can resemble *R. caesius* very closely.

## RUBUS FRUTICOSUS L. agg.

The aggregate comprises all Brambles with the exception of *R. caesius* L. Determination is a matter for the expert, and keys have been supplied not in the first instance for recording purposes, but that the interested person may come in time to know his local brambles.

A new key to groups is given below, and it is worthwhile attempting to place plants in the appropriate group. The species key was compiled in 1976 for use in Durham. Unfortunately, the key could not be expanded to take in all the Rubi of the region, because up-to-date information has not been available.

Voucher specimens are essential, and should comprise both of the following:

a) a piece of stem from the middle of the first-years growth, with two good, typical leaves attached;

b) the whole inflorescence with at least buds and flowers, and a leaf below the inflorescence.

Care should be taken to collect these from the same bush, and a note made if the bush is at all shaded. Note also the colour of the petals.

### Key to Sections and Sub-sections

The groups are, for the most part, artificial, and there is no general agreement on the sub-division of the genus. The 1978 key given below divides the genus into 7 main sections, and 7 sub-sections of *Appendiculati*.

1. Stems usually trailing; leaves mostly ternate, or lowest leaflets sessile; lateral leaflets ± sessile; flowers with large, roundish petals; fruit usually imperfect, of relatively few large, dull-bloomed drupelets; leaves not deciduous in winter. TRIVIALES

   Not as above. 2

2. Stems ± erect (like raspberry canes), not rooting at the tip; panicles simple-racemose. SUBERECTI

   Stems high- or low-arching; often rooting at tip. 3

3. Stems (1st year) without stalked glands or pricklets. 4

Stems (1st year) with stalked glands and/or pricklets. APPENDICULATI 7

4. Panicles eglandular, or with sparse, stalked glands. Stem not densely hairy. 5

Panicles with some stalked glands; stem ± densely hairy. Vestitae

5. Leaflets chalky-white beneath; stem strongly pruinose. R. ulmifolius & DISCOLORES

Leaflets whitish or grey beneath. 6

6. Stamens shorter than styles; stems low-arching. SPRENGELIANI

Stamens equalling, or longer than the styles. Stems various. SYLVATICI

7. Stems with pricklets and stalked glands short, well-differentiated from the main prickles. 8

Stems with prickles, pricklets, stalked glands, and other armature, of various sizes. 9

8. Terminal leaflet mucronate to cuspidate. Mucronati

Terminal leaflet not mucronate or cuspidate. Radulae and Apiculati (and others in the S. of England)

9. Stalked glands rare (as distinct from gland-tipped prickles, acicles etc). Anisacanthi

Stalked glands frequent. 10

10. Armature stout; stems strong, high-arching. Hystrices

Armature slender; stems weak, low-arching. Glandulosi

A. Newton, pers. comm. (1978)

Key to species in v.c.66

1. Leaves deeply lacinate (an escape from cultivation). R. laciniatus Willd.
Leaves serrate, not deeply lacinate. 2

2. Stems usually trailing; leaves mostly ternate or lowest leaflets sessile or sub-sessile; stipules lanceolate; fruit usually imperfect, of few relatively large drupelets. (Sect. Triviales) 3
Not as above. 4

3. a) Stems robust with crowded, mostly patent, strong, unequal prickles; glands ± numerous. Leaves (3)5-nate, a good deal strigose above, thickly greyish-pubescent to felted beneath; terminal leaflet widely cordate; sharply and unequally incised; perianth broad, dense; pedicels villose and felted, with numerous submerged glands. Flowers large, petals white or pinkish. R. dumetorum agg. (incl. R. tuberculatus Bab.)

b) Stems sharply angled; prickles short, patent, nearly equal; stalked glands (if present) few and short. Leaves often 4-5-nate or sub-compound; leaflets deeply, coarsely and bluntly serrate, white-felted and velvety-pubescent beneath; basal leaflets sub-sessile and large imbricate, terminal rhomboidal. Flowers pink, sepals reflexed. R. eboracensis W.C.R. Wats.

c) Not as above. un-named Triviales, or recent hybrids, or R. caesius L.

4. Barren stems ± erect (like raspberry canes), never rooting at the tip. (Sect. Suberecti) 5
Stems high- or low-arching, rooting at tips (when allowed) 6

5. a) Stems erect, tall, glabrous-glaucescent; lower prickles reduced to pricklets, purplish-black. Leaves large, 5(-7)nate, thin, shining green on both sides. Petals white, glabrous; stamens longer than the styles; carpels and receptacle glabrous; ripe fruit dark-red. Apparently rare. R. nessensis Hall

b) Stems sub-erect, short, prickles numerous, subulate, crowded. Leaves 6-7-nate; petioles distinctly furrowed throughout; leaflets thick, rugose, densely pubescent beneath, imbricate, plicate. Stamens equalling or shorter than styles; carpels and receptacle pilose; fruit dark-red. R. scissus W.C.R. Wats.

c) Prickles slender, but longer and stronger than in the two preceding species, falcate, yellow and red. Leaves 3-5-nate; leaflets imbricate, terminal leaflet broad, plicate. Sepals cuspidate, receptacle strongly hairy. Petals white or pink, abruptly clawed, hairy. Fruit black. Apparently rare. R. plicatus Weihe & Nees

d) Prickles longer; leaves neater, more evenly serrate and acuminate, not imbricate; ripe fruit purplish-black. R. fissus Lindley

6. Barren stems ± eglandular, pricklets absent or very sparse. 7
Barren stems glandular and/or with pricklets. 15

7. Panicles with numerous stalked glands, stems pilose. 8
Panicles eglandular (or with a few minute glands) 11

8. Terminal leaflet roundish or roundish-obovate, mostly as broad as long. The whole plant with a purplish or a brownish tinge. 9
Terminal leaflet ovate-elliptical, mostly longer than broad. The whole plant with a greyish tinge. 10

9. Plant suffused purplish in open sites, with a preponderance of patent stem prickles. Leaflets mostly 5, dull above, strongly pectinate-pilose on the veins beneath, densely greenish-tomentose beneath. Panicle long, hardly narrowed to abrupt apex. Petals white to deep rose-pink. R. vestitus Weihe & Nees
Plant suffused brownish in open sites, with strongly declining prickles, and a few falcate with stout bases. Leaves thick, often 3-5-nate, lobate, always pedate, sparsely hairy beneath. Panicle long, slightly flexuous, pyramidal. Sepals glandular-aciculate; anthers glabrous, or with a very few hairs. Petals pale-pink. R. anisacanthus G. Braun

10. Stems angled, prickles declining. Leaves plicate, undulate, densely greyish-tomentose beneath. Anthers usually pilose; stamens exceeding the yellowish styles. R. criniger (E.F. Linton) Rogers
Stem roundish, not furrowed; pricklets slender, rather few. Leaflets 3-5-nate, imbricate, pubescent or thinly-hairy, and greenish-felted beneath. Anthers hairy, pink, much exceeding the pink styles. Petals pink. R. mucronulatus Bor.

11. Petals white. 12
Petals pink. 13

12. Small, neat leaflets, grey-felted beneath; terminal leaflet small and obvate, long-stalked. Panicle with many strong curved prickles. R. lindebergii P.J. Muell.
Stems shining, with large strong prickles. Leaflets sharply toothed, green beneath; terminal leaflet elliptic-obovate, usually cuneate-based (wedge-shaped). Panicle broad and intricately branched. R. lindleianus Lees

13. Leaflets 3-5, small, dark-green, somewhat rugose above, chalky-white-tomentose beneath. Prickles strong-falcate. Stems pruinose-waxy in the open. Not on acid soils or high ground. R. ulmifolius Schott
Leaflets green- or grey-felted beneath. 14

14. a) Panicle strongly armed with hooked prickles. Stamens ± equal to, or shorter than styles. Petals notched. R. nemoralis P.J. Muell. (R. selmeri Lindb.)

b) Leaves light dull-green in the open with convex, nearly simply-serrate leaflets, occasionally 6-7-nate. Panicle long, with numerous long, slanting, reddish prickles (and rare glands). Stamens longer than the greenish or rosy-based styles. R. polyanthemus Lindeb.

c) Stamens pink, longer than the yellow styles; anthers pilose, reddish; leaflets broadly imbricate, pubescent beneath; terminal leaflet various; roundish, sub-quadrate apiculate, or obovate. Panicle with long-stalked lower leaves. R. errabundus W.C.R. Wats.

15. Stems with pricklets and glands short. 16
Stems with pricklets of various sizes. 18

16. Stems glabrous, rachis glabrous and acicular below. Leaflets coarsely and jaggedly compound-serrate. Overall cast purplish. Petals white, notched. R. echinatoides (Rogers) Sudre

Stem moderately hairy, and/or rachis hairy. Leaflets neater and more finely-serrate. Overall cast reddish. Petals pink or white. 17

17. Leaflets roundish-obovate, small, pedate. Panicles wide-spreading with a number of stout falcate prickles. Stamens deep pink ± equal to the styles. Flowers cup-shaped, petals pale pink. R. infestus Weihe & Nees (R. taeniarum)

Leaflets rather long, ovate, digitate, grey or grey-white beneath. Panicles pyramidal. Stem with many small tubercularpricklets; stamens white, longer than the styles. Petals usually pink. R. radula Weihe ex Boenn

18. Plant greyish. Leaflets rugose, often convex, almost imbricate. Terminal leaflet rhomboidal. Stem with light-brown, yellow-tipped slanted prickles; glands short-stalked. Panicles compound. Sepals with short pricklets, at first loosely reflexed, then loosely clasping the fruit. R. adenanthoides A. Newton

Leaves thick, terminal leaflet long-stalked, elliptic-obovate, cuspidate. Stem hairy with some hooked prickles and long glands, acicles and pricklets. Panicles simple. Sepals without pricklets, at first patent, ultimately sharply-reflexed. R. dasyphyllus Rogers

mainly A. Newton, pers. comm, 1974, 1978.

POTENTILLA ANGLICA Laisch./P. ERECTA (L.) Rausch./P. REPTANS L.

Identification of these species is made more difficult by the presence of hybrids, two of which may be widespread. True P. anglica has been much over-recorded in the region in the past.

P. x mixta Nolte ex Reichb.

Experimental hybridisation has revealed two possible origins for P. x mixta, i) between P. reptans L. and P. anglica Laisch., and ii) between P. reptans and P. erecta (L.) Rausch.

The hybrid forms runners, but is distinguished from P. reptans by the presence of 3- and 4- as well as 5-nate leaves, by the 4- and 5-merous flowers, and by its sterility. Anthers have (usually) 0-10% viable pollen, and achene formation is rare. A succession of empty calices along the stem is a good indication of the hybrid.

P. x suberecta Zimm. (P. anglica Laich. x P. erecta (L.) Rausch.). The hybrid plant is intermediate between the parents in many characters which include leaflet- and petal-number, petiole-length, and flower-diameter. The suberect stems resemble those of P. erecta, but they root at the nodes in late summer, as do those of P. anglica. Pollen viability varies from 0-40%, and a few achenes are usually produced (but rarely more than 10 per flower.

B. Matfield, in Stace (1975)

| | P. erecta | P.reptans | P. x mixta | P. anglica | P. x suberecta |
|---|---|---|---|---|---|
| stem | not rooting | rooting | often rooting | rooting late | sometimes rooting |
| stipules | deeply cut | entire, broad | entire, narrow | entire or 3-fid | entire, cleft |
| petioles | 0 or very short | long | short to medium | medium | short |
| leaflets | oblanceolate | obovate | obovate to oblanceolate | obovate | narrow |
| -upper | 3 | 5 | 3-5 | 3(5) | 3 |
| -lower | 3(5) | 5(7) | 5 | 5(3) | 3 |
| -serrations | acute | obtuse | subacute | acute | acute |
| flower diam (mm) | 11 | 22 | 14-16 | 15 | 18 |
| petal number | 4(5) | 5 | 4(5) | 4(5) | 4(5) |
| ripe carpels | reticulate | granulate | granulate | tuberculate or reticulate | reticulate |

Grose, Flora of Wiltshire (1957)

GEUM x INTERMEDIUM Ehrh. (G. urbanum L. x G. rivale L.)

This hybrid appears quite widespread in the region and usually occurs with the parents. It is intermediate in flower colour and size, stipule size, etc., but is extremely variable. It is fertile and hybrid swarms are found.

AGRIMONIA L.

When well-developed, the two British species may be separated as follows:

Leaves not (or slightly) glandular beneath; fruit hypanthium (receptacle) obconic, deeply grooved throughout, basal spines spreading laterally. The common species in the region. A. eupatoria L.

Very robust and sweet-smelling; leaves with numerous sessile glands beneath; fruit hypanthium campanulate, without grooves at the base, basal spines deflexed. Mostly in association with acid woodland but may occur in calcareous soils if wet enough. Rare in the region. A. procera Wallr.
(A. odorata (Gouan)Mill.)

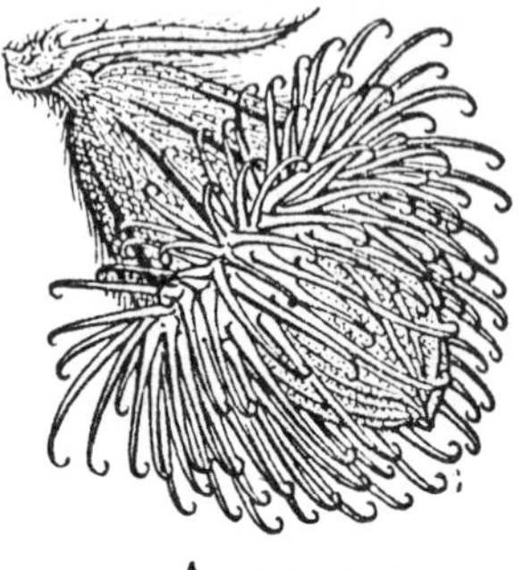

Fruit of A. eupatoria A. procera

Problems of identification can arise with:

i) forms of A. procera in which a single seed is formed instead of the normal two, when the hypanthium loses its characteristic shape.

ii) large, glandular, distinctly aromatic forms of A. eupatoria, which can be mistaken for A. procera; this form has been noted in Durham - on the limestone at Cassop Vale, and in Hawthorn Dene.

The hybrid A. eupatoria x A. procera was found in the 1940's in v.c.67. No fruits are formed.

ALCHEMILLA VULGARIS L.

The genus Alchemilla is well-represented in our region, particularly in Teesdale. Three species, A. acutiloba, A. monticola, and A. subcrenata are endemic in Durham. Species are generally clear-cut, but careful attention to detail and the characters as a whole is necessary in this critical genus. It is best to become thoroughly aquainted with the common lowland species first, when the more local upland species will be the more easily detected in the field. Mature plants with well-developed leaves and inflorescences should be examined. Plants with immature leaves can be impossible to name.

1. Stems and petioles with spreading (erecto-patent, patent, deflexed) hairs.

2. Urceole (= receptacle in CTW, hypanthium in Fl.Eur.) ± densely hairy.

3. All pedicels ± densely hairy.

4. Leaves ± orbicular; leaf lobes with teeth few (4-6) and relatively large; whole plant silky-hairy; base of petiole (as distinct from stipules) not reddish. Not found in our region. A. glaucescens Wallr.

4. Leaves ± reniform, sinus open; leaf-lobes with 6-9 teeth; whole plant not silky-hairy; base of petiole (as distinct from stipules) reddish. A. filicaulis Buser subsp. vestita (Buser) M.E.Bradshaw

3. At least some pedicels glabrous or sub-glabrous.

5. Dwarf plants, up to 15 cm.

6. Leaf-lobes with distinct incisions; base of petiole (as distinct from stipule) not reddish. Not found in our area. A. minima Walters

6. Leaf-lobes without distinct incisions; base of petiole wine-red. (A. filicaulis)

5. Medium-sized plants, up to 50cm.

7. Leaves ± orbicular, densely and evenly hairy above; lobes with even teeth, and usually with a distinct, short incision between the lobes. A. monticola Opiz

7. Leaves ± reniform, often sparsely or unevenly hairy above; lobes restricted, with unequal teeth. (A. filicaulis)

2. Urceole glabrous, or sparsely hairy.

8. Epicalyx segments at least as long as the sepals; mature achene much longer than the urceole; all parts densely hairy. A. mollis (Buser) Rothm.

8. Epicalyx segments usually shorter than the sepals; mature achene as long as the urceole.

9. Dwarf plants, up to 15 cm.

10. Leaf-lobes with distinct incisions; base of petioles not reddish. A. minima Walters

10. Leaf-lobes without distinct incisions; base of petioles reddish. (A. filicaulis)

9. Medium-sized plants, to 50cm.

11. Leaves glabrous or subglabrous above; flowers small, to 3mm wide. A. xanthochlora Rothm.

11. Leaves hairy above, at least on the folds; flowers usually larger.

12. Hairs on stems and petioles all patent or erecto-patent; leaf shape various.

13. Hairs erecto-patent; urceole attenuate at base; inflorescence lax; lower part of petioles pinkish. A. gracilis Opiz

13. Hairs on stems and petioles patent; urceole rounded at base; inflorescence not lax; base of petioles not pinkish.

14. All leaves densely and evenly hairy on both surfaces; leaves orbicular; leaf lobes with equal teeth. A. monticola Opiz

14. Some leaves sparsely or unevenly hairy above.

15. Leaf-lobes almost triangular, often long; base of petioles not reddish; urceole glabrous. A. acutiloba Opiz

15. Leaf-lobes ± rounded; base of petioles reddish; urceole often somewhat hairy.

16. Upper part of stem (above the 4th internode), inflorescence-branches, and pedicels glabrous; lower leaf surface (except veins) glabrous or nearly so. A. filicaulis subsp. filicaulis

16. Hairy throughout, but density variable. A. filicaulis subsp. vestita

12. At least some hairs on stems and petioles deflexed; leaves orbicular or sub-orbicular.

17. Leaves undulate, rather sparsely hairy above, often only on folds; basal lobes turned up and touching in the live plant. A. subcrenata Buser

17. Leaves densely and evenly hairy above; flowers very small, 1.5-2.5mm wide. Not found in our area. A. tytthantha Juz.

1. Stems and petioles with appressed or sub-appressed hairs.

18. Epicalyx segments usually at least as long as the sepals; mature achenes much longer than the urceole. Cultivated plant. Not found in the region. (not generally accepted name f(no generally accepted name for this taxon)

18. Epicalyx segments usually much shorter than the sepals; mature achenes as long as the urceole.

19. Stem and petioles densely clothed with silky hairs; leaves undulate (folded when pressed), distinctly hairy above; sinus closed; flowers densely-clustered; flowers glabrous. A. glomerulans Buser

19. Stems and petioles nearly glabrous (hairy only below); leaves glabrous or nearly so; sinus open or closed; flower clusters lax.

20. Leaves ± reniform with open basal sinus, sometimes subglabrous above; without distinct incision between lobes; teeth few, broad, unequal, not connivent. A. glabra Neygenf.

20. Leaves orbicular-reniform, glabrous above; with incision between lobes; teeth numerous, narrow, equal, connivent (curved towards apex of lobe). A. wichurae (Buser) Stefánsson

M.E. Bradshaw, pers.comm. (1979)

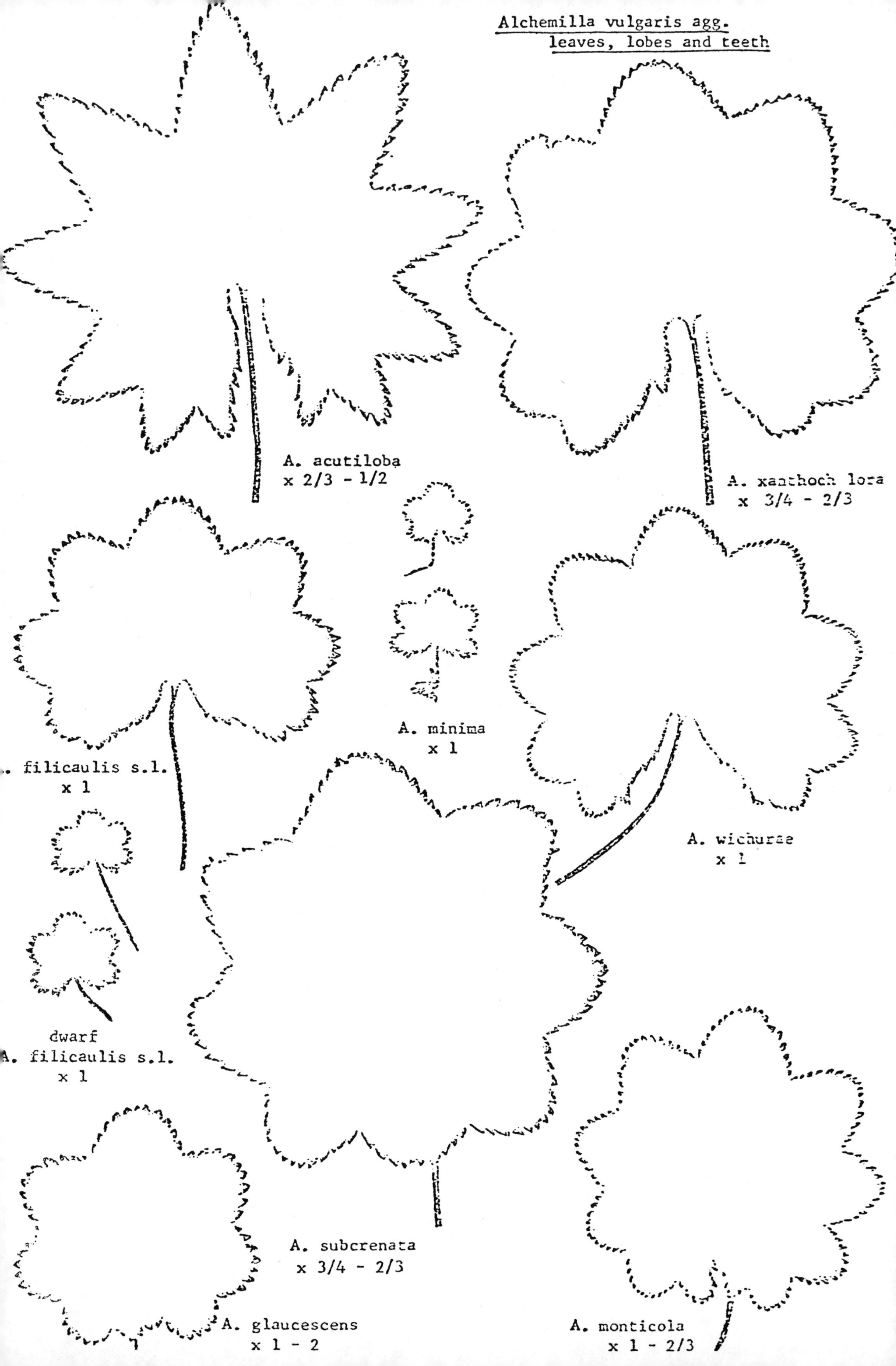

Alchemilla vulgaris agg.
leaves, lobes and teeth
A. acutiloba
x 2/3 - 1/2
A. xanthochlora
x 3/4 - 2/3
A. minima
x 1
. filicaulis s.l.
x 1
A. wichurae
x 1
dwarf
A. filicaulis s.l.
x 1
A. subcrenata
x 3/4 - 2/3
A. glaucescens
x 1 - 2
A. monticola
x 1 - 2/3

Part of inflorescence of A. vulgaris

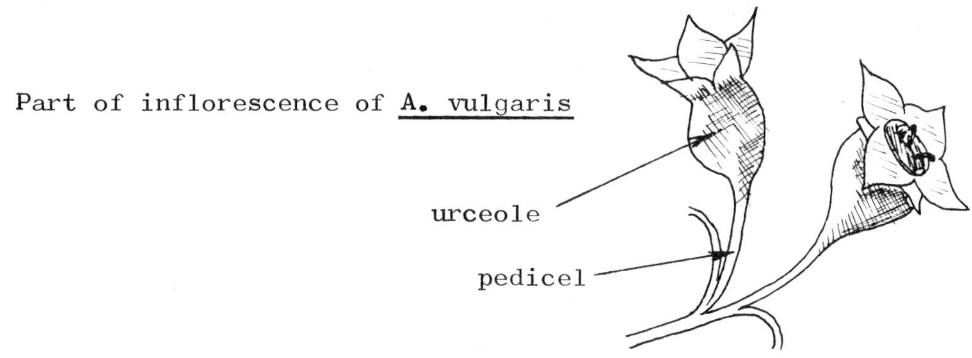


APHANES L.

Lobes of stipules surrounding inflorescence triangular-ovate; fruit 2.2-2.6mm, including the divergent sepals. A. arvensis L.

Lobes of stipules oblong; fruit 1.4-1.8mm, including the convergent sepals. A. microcarpa (Boiss. & Reut.) Rothm.

The difference in overall colour mentioned in CTW is not reliable. A. microcarpa is scarce, occurring mainly on the more acidic, sandy soils.

S.M. Walters, in Flora Europaea Vol.2(1968)

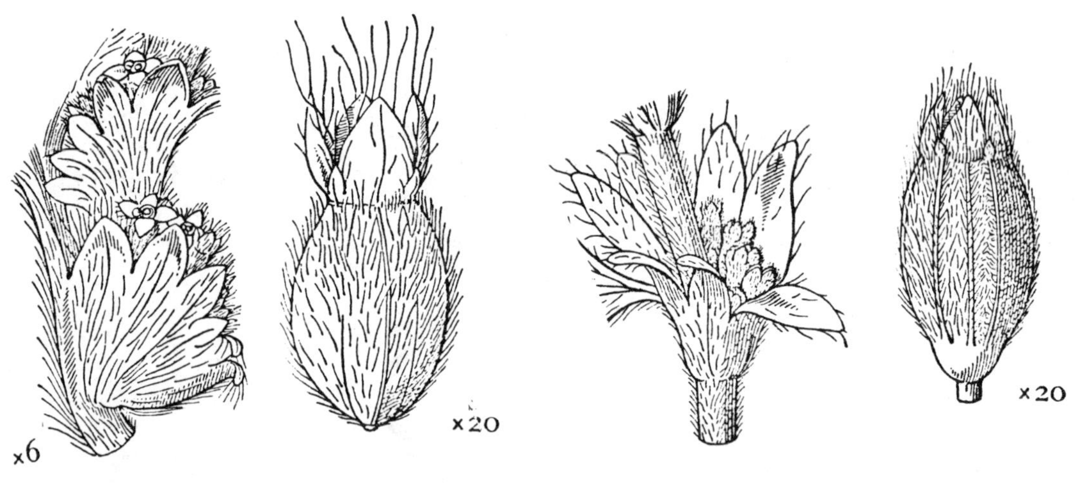


A. arvensis A. microcarpa

## ROSA L.

The genus Rosa is particularly complex and poses special problems for the recorder. Because of their propensity for hybridisation and their peculiar breeding system, roses now present the recorder with a complex system of taxa. These taxa were in the 19th century afforded species level, but in the early part of this century, common sense prevailed, and most described taxa were relegated to varietal level under (in Britain) a dozen or so species. Wolley-Dod's work in the 1920's and 1930's was a noble (and still relevant) attempt to put order into chaos.

Unfortunately the extent of hybridisation in the genus was at first recognised by a mere handful of rhodologists, and most field-workers were content to use Wolly-Dod's monograph (J. Botany Suppl., 1930, 1931) rather too closely and uncritically. In this work many varieties are listed as having affinities with species A when in reality they are complex hybrids involving species B and C. Consequently, species boundaries were widened and blurred, and no diagnostic characters could be adduced for separating one species from another.

Although it is now possible, we feel, to delimit rose species more accurately, the field worker must recognise the fact that in many areas, especially where the eu-caninae overlap, hybrids and complex hybrids may greatly outnumber the true species. Only those bushes which agree fully with the species diagnosis should be recorded. The rest should be ignored until experience has been gained. As with many other critical groups of plants, a few hours in the field with a knowledgeable person is almost a prerequisite to the study of roses. The following key is thus offered mainly to those who have had some experience in this genus.

Roses are best studied in late summer when fruit is beginning to form; varieties and hybrids can only be satisfactorily elucidated when the fruit is fully formed. The rose season, therefore, runs from late August to early October with a few species discernible in July.

### Main features for the identification of roses

| | |
|---|---|
| styles | whether columnar or not; whether glabrous, pilose (hairy), or villous (woolly); slightly-exserted or in a dense hemispherical head sessile on the disc. |
| stylar aperture | width in relation to disc |
| disc | flat, or in a low or high pyramid |
| fruit | hispid or smooth; shape especially below the disc |
| sepals | position in the fruiting stage i.e., erect, spreading or reflexed; deciduous or persistent; hispid or smooth; simple or pinnate |
| pedicels | short (c.1cm) medium (2cm) or long (3cm +); hispid or smooth; few or many |
| leaves | shape; serration; size; presence or absence of pubescence and/or glands; type of glands - odourless, aromatic or with fruity (apple-scented) odour when crushed |
| prickles | straight, curved or hooked; stout-based or slender; equal or unequal; mixed with acicles or not; mixed with glands on pedicels or not |
| habit | tall or short; strong vigorous bush or more slender; branches or stems straight, zig-zag or arching; height (c.1m, c.1.5-2m, over 2m) |

The main categories and species of British Roses

field-roses — *R.arvensis* (reaching the southern part of our region) *R.stylosa* is strictly speaking a member of the *caninae*, but conveniently placed in the field-rose group as, like *R.arvensis*, its styles are united into a column (far south of Britain only)

dog-rose — *R.canina* (very common), *R.dumetorum* (common), *R.afzeliana* (very common), *R.coriifolia* (very common), *R.obtusifolia* (rare in the region)
*R.afzeliana* Group *Subcaninae* (this hybrid group is in some areas more prevalent than the true species). *R.coriifolia* Group *subcollinae* (a hybrid group with stable forms, again, at least in parts of the region, more common than the parent species)

downy-rose — *R.tomentosa* (rare in the region), *R.sherardii* (common), *R.mollis* (very common)

sweet-briar — *R.rubiginosa* (uncommon in the region); *R.micrantha* (rare in the region); *R.agrestis* (very rare nationally - not now in the region)

burnet rose — [*R.pimpinellifolia* (not as common now through loss of habitats;] mainly on the coast)

KEY TO British Rose species

1. Styles exserted and united into a column.

2. Weak scrambling shrub, younger stems dark-green or glaucous, often purple-tinged; petals chalky-white; stylar-column ± fused and slender, as long as the stamens, on a flat disc; sepals <1cm, ovate-acuminate, the tip not expanded, lobes few, soon falling; pedicels long, 2-5cm, normally with stalked glands. *R. arvensis* Huds.

2. Strong arching shrub 1-4m; petals pink or white; stylar-column glabrous, plump, agglutinated when young, but becoming free, shorter than the stamens, on a prominently conical disc; stipules and bracts rather narrow; pedicels long, 2-4cm, with (usually) stalked glands. *R. stylosa* Desv.

1. Styles not united into a column although sometimes long-exserted, (often apparently so in herbarium specimens where the fruit has shrunk).

3. Low shrub with erect stems, spreading vegetatively and forming large patches. Flowers solitary, without bracts; sepals entire, erect and persistent; leaflets small 0.5-1.5cm, serration simple, (3-)4-5 pairs; stems densely prickly and bristly with acicles; fruit purplish-black, subglobose. *R. pimpinellifolia* L. (*R. spinosissima* auct.)

3. Flowers with 1 or more bracts, serration simple or double, leaflets 2-3 (-4) pairs, stems with stout prickles, rarely unarmed, or if with dense acicles, then leaves with fruity fragrance.

4. Leaflets glabrous, sometimes with a few reddish, stalked, non-odorous glands along petiole and midrib.

5. Styles glabrous, hispid or rarely woolly, forming a ± loose group, orifice of disc about 1/5 its width. *R. canina* L.

leaflets simply serrate - Group Lutetianae
leaflets irregularly serrate - Group Transitoriae
leaflets doubly serrate - Group Dumales

plants mainly caninae but with glandular-hispid peduncles and fruits are hybrids. Group Andegavensis

plants mainly caninae, but leaflets glandular-biserrate; and with glands on petioles and usually on peduncles. Group Scabratae

5. Styles forming a dense, woolly, hemispherical mass almost covering the disc; orifice of disc about 1/3 of its width; pedicels short, 0.5-1.5cm ; sepals spreading-erect after flowering, finally ascending-erect, falling late; illuminated parts of bush often with strong anthocyanin colouration, contrasting with the glaucous-green leaves. R. afzeliana Fr.

forms with spreading-reflexed sepals, styles in a more rounded head, and other diminished R. afzeliana traits are hybrids. Group Subcaninae

4. Leaflets tomentose or hairy below, at least on the midrib.

6. Leaflets with prominent viscous brownish or translucent glands on the lower surface, giving a sharp, fruity (? apple) sweet-briar odour when rubbed.

7. Leaflets cuneate at base, pedicels glabrous. R. agrestis Savi

7. Leaflets rounded at the base, pedicels glandular-hispid.

8. Stems erect; prickles unequal; primary stems often ± densely acicular; styles hispid; sepals erect or spreading, persistent at least until the fruit reddens. R. rubiginosa L.

8. Stems lax, tall, arching; prickles stout, uniform; leaves longer and usually narrower in proportion to length, and more acute at apex; subfoliar glands fewer and smaller; peduncles longer; sepals reflexed and deciduous earlier, often before fruit reddens; styles nearly always glabrous; fruit often contracted below the disc. R. micrantha Borrer ex Sm.

6. Leaflets glandless, or with ± inconspicuous glands which may produce an aromatic (turpentine-like) odour when rubbed.

9. Leaflets hairy or tomentose, glandless (rarely a few non-odorous glands on the midrib), simply or biserrate.

10. Styles glabrous, hispid (rarely woolly), sometimes slightly exserted on a low, conical disc; orifice of disc about 1/5 of its width; sepals reflexed, and soon falling; habit very strong, tall, and arching, 2-3m or more. R. dumetorum Thuill.

parallel hybrid forms to those of R. canina are found.
leaflets simply or biserrate, without subfoliar glands, peduncles glandular-hispid. Group Deseglisei
leaflets biserrate with subfoliar glands, peduncles glandular-hispid. Group Mercicae

10. Styles woolly in a dense, hemispherical mass almost covering the disc; orifice of disc about 1/3 its width; pedicels short, 0.5-1.5cm ; sepals spreading-erect after flowering, finally ascending-erect, falling late; anthocyanin pigmentation as in

<u>R. afzeliana</u>; habit very strong, with stout prickles, but more of a bush than <u>R. dumetorum</u>. Flowers often in large clusters. <u>R. coriifolia</u> Fr.

[forms with spreading-reflexed sepals, falling sooner; styles in a more rounded head, and other diminished <u>R. coriifolia</u> traits are hybrids. They are often very difficult to distinguish from either <u>R. dumetorum</u> or from <u>R. obtusifolia</u>. Group <u>Subcollinae</u>]

9. Leaflets usually glandular-biserrate, usually with some glands below at least on the midrib.

11. Orifice of disc about 1/5 of its width.

12. Stem with stout, very strongly-hooked prickles; leaflets with a rounded base; in most varieties smaller and neater, and certainly rounder than in 10; sepals short, broad, and pinnate, reflexed, usually not glandular on the back; pedicels short, 0.5-1.5cm. <u>R. obtusifolia</u> Desv.

12. Stem prickles more slender than in <u>R. obtusifolia</u>, but stronger than in <u>R. sherardii</u> or <u>R. mollis</u>, arching to ± straight; sepals elongated, not strongly pinnate, glandular-setose on back, erect or spreading, but markedly constricted at the attachment, falling before the fruit ripens. Styles pilose or glabrous. Leaflets rather harshly-tomentose. Pedicels long, glandular. Growth moderately strong and arching, 1.5-2m. <u>R. tomentosa</u> Sm.

[hybrid forms occur with reflexed sepals, less tomentose leaflets, styles ± glabrous. Group <u>Scabriusculae</u>]

11. Orifice of the disc about 1/3 its width or more; stipules very broad, and hiding the short peduncles; leaflets densely and softly-tomentose; prickles more slender; stigmas densely villous in a ± broad, flat head.

13. Sepals fully erect, ± simple, the tapering tips expanded into a small blade, ± fused to the fruit and persisting until it falls or decays; orifice of disc over 1/3 and often 1/2 the width; prickles slender, quite straight, subulate, their bases little thickened; low erect shrub 0.5-1.5m, with very straight main stems and branches. <u>R. mollis</u> Sm. (<u>R. villosa</u> L.)

13. Sepals erect, or spreading-erect, generally with small lateral pinnae, slightly constricted at the attachment, persisting until the fruit ripens but finally falling; orifice about 1/3 the width of the disc; prickles slender, inclined or slightly curved at least on some parts of the bush; slender, arching shrub c. 1.5m, with flexuous stems. (beware of lopped bushes which may appear upright). <u>R. sherardii</u> Davies

R. Melville (1970): suggestions for northern roses G.G. Graham (1973): emended R. Melville (1974): emended G.G. Graham (1976, 1980).

Neither <u>R. agrestis</u> nor <u>R. stylosa</u> is recorded in the region.

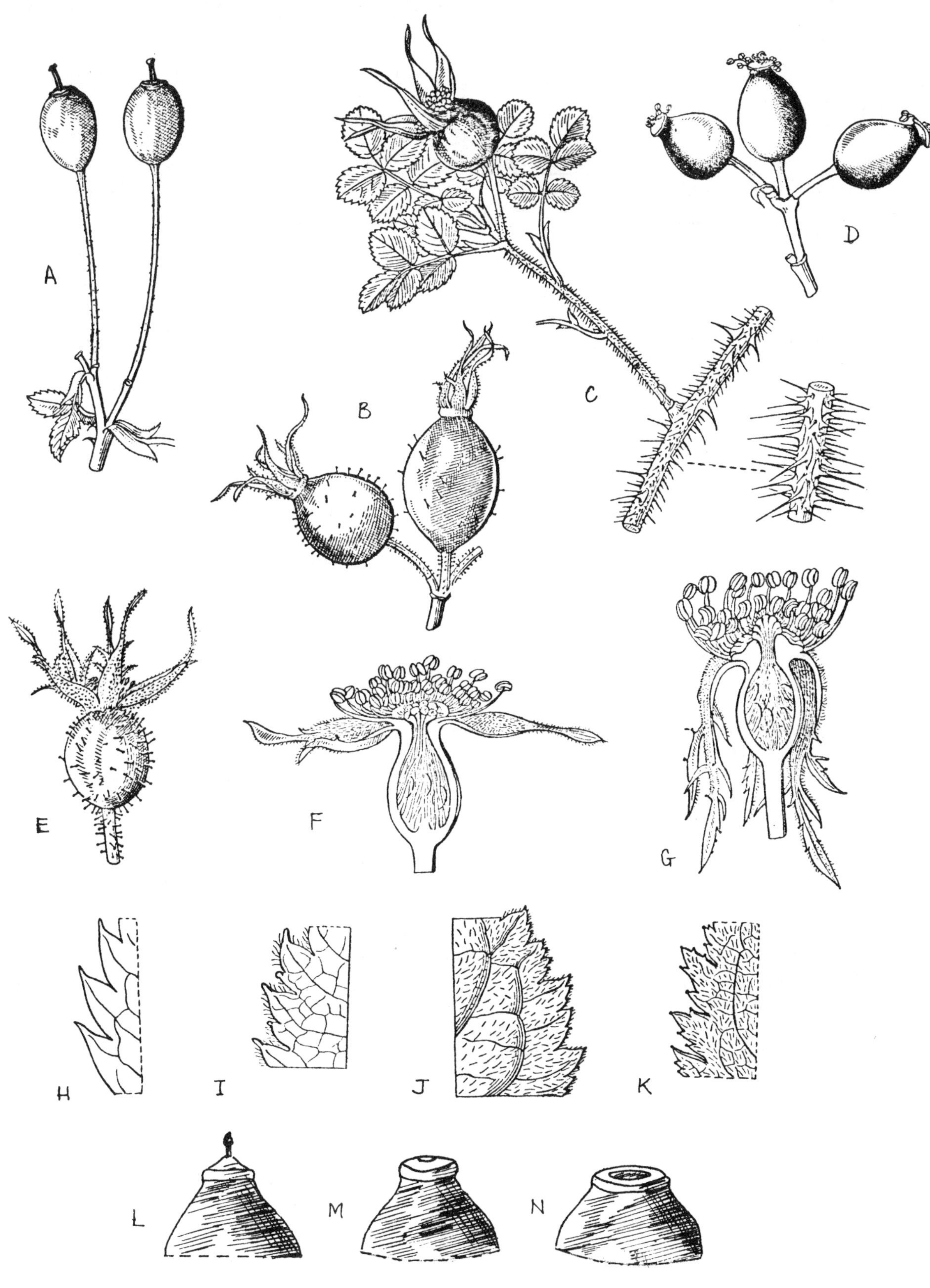


A Fruit of R. arvensis
B Fruit of R. mollis, sepals mostly fully erect
C R. pimpinellifolia
D Fruit on short pedicels (e.g. R. canina, R. coriifolia)
E Fruit of R. sherardii, sepals ascending
F Sepals patent in fruit, e.g. R. tomentosa
G Sepals reflexed in fruit, e.g. R. canina
H Leaf toothing simple
I Leaf biserrate
JK Leaf multiple-serrate
L Disc conical, styles in short column, R. stylosa
M Orifice small, e.g. R. canina
N Orifice large, e.g. R. mollis

PRUNUS L.

Two introduced species, P. cerasus and P. cerasifera occur in hedges in the region, and though not common, are doubtless overlooked. In addition, the evergreen P. laurocerasus and P. lusitanica are sometimes planted.

1. Flowers in long, loose racemes; leaves folded in bud, closely- and sharply-serrate, glabrous except sometimes for axial tufts of hair beneath; petiole with a ± opposite gland on each side just below the lamina. P. padus L.

1. Flowers not in racemes; leaves convolute or folded in bud, serrations various; glabrous or hairy; petiole with or without glands.

2. Leaves convolute (rolled) in bud; shoots with no terminal bud; bud-scales caducous, not forming an involucre around the inflorescence; pedicel shorter than or only slightly longer than the ripe fruit.

3. Young twigs dull, usually pubescent; young leaves pubescent, becoming glabrous and dull above; flowers not solitary.

4. Usually a shrub, often somewhat thorny; young twigs and pedicels conspicuously pubescent; petals pure white.
P. domestica L. subsp. insititia (L.) C.K.Schneid.

4. Shrub or small tree, never thorny; twigs sparingly pubescent at first then glabrous; pedicels usually glabrous; petals tinged with green, especially in bud. P. domestica subsp. domestica

3. Twigs glabrous from the first, usually remaining green in the second year, rather glossy; leaves somewhat glossy above; flowers usually solitary. P. cerasifera Ehrh.

2. Leaves folded in bud; shoots with a terminal bud; bud-scales persistent, forming an involucre around the inflorescence; pedicel at least twice as long as ripe fruit.

5. Usually a shrub (rarely a small tree to 5m ); leaves dark-green, shining above, soon glabrous beneath; petiole 1-3 cm; hypanthium not constricted at apex (campanulate). P. cerasus L.

5. Tree to 25m high; leaves dull above, usually with some persistent pubescence beneath; petiole 2-5cm; hypanthium constricted at apex (urceolate). P. avium L.

CRATAEGUS L.

C. laevigata (Poir.) DC. (C. oxyacanthoides Thuill.) has been reported in Durham and Cumbria - occurring occasionally in hedges and wood-edges. C. laevigata is a southern plant, and it seems very unlikely to occur in our region. These reports are much more likely to be of the hybrid C. monogyna x C. laevigata (C. x media Bechst.). Hybrids are intermediate between the parents in all characters, of which leaf shape (primarily the degree of indentation) is the most diagnostic.

A.D. Bradshaw, in Stace, (1975).

SORBUS L.

Of the members of this genus occurring in our region, only S. aucuparia, S. rupicola, and S. lancastriensis are native plants. S. rupicola occurs in a few places in Teesdale and Cumberland, whilst S. lancastriensis is endemic to several localities around Morecambe Bay, both species on limestone. Identification often depends upon critical examination of leaf characters, and several leaves from each plant should be examined.

1. Leaves pinnate, with 4 or more pairs of leaflets. S. aucuparia L.
   Leaves not pinnate. .. 2

2. Large trees. Leaves green on both sides, subglabrous beneath except when very young, deeply lobed, lowest lobe often extending 1/2 way to midrib, lobes acuminate, leaves finely often doubly serrate, with 4-6 pairs of veins; fruit brown, much longer than broad. S. torminalis (L.)Crantz.
   Leaves felted beneath, with more than 6 pairs of veins. .. 3

3. Tree. Leaves finely-tomentose beneath, with triangular, acute or acuminate lobes extending 1/7 to 1/4 the way to the midrib, with (6)7-10(12) pairs of veins; fruit orange-brown to brown, subglobose or longer than broad. Planted. S. x latifolia (Lam.)Pers. (S.aria x S. torminalis)
   Leaves without acute or acuminate triangular lobes; fruit red. .. 4

4. Tree. Leaves yellowish-grey tomentose beneath, with broad, ascending, serrate lobes extending 1/4 to 1/3 the way to the midrib, the lowest occasionally a free leaflet on shade leaves, with 7-9 pairs of veins; fruit much longer than broad, scarlet. S. intermedia (Ehrh.) Pers.
   Leaves unlobed, or very shallowly lobed, grey-white tomentose beneath. .. 5

5. Shrub to small tree. Leaves ovate, elliptical or oval, very rarely obovate, doubly crenate-serrate or shallowly lobed, entire in basal 1/5 or less, with (9)10-14(16) pairs of veins; fruit longer than broad (rarely globose). Very variable in many characters. S. aria (L.)Crantz.
   Shrub or small tree. Leaves obovate with cuneate base, not lobed; entire at least in lower 1/4; fruit broader than long. .. 6

6. Shrub or small tree. Leaves obovate, ± entire in basal 1/3 to 1/2, tapering evenly to a cuneate base for at least the lowest 1/3 of the leaf, leaves up to twice or more as long as broad, with (6)7-9(10) pairs of veins. Exposed upland and mountain limestone. S. rupicola (Syme)Hedl.
   Large shrub. Leaves obovate, entire or nearly so in basal 1/3, tapering to an always cuneate base for about 1/4 of the distance above the base, leaves seldom more than 1.7 times as long as broad, with (7)8-10(12) pairs of veins. S. lancastriensis E.F. Warb.

A. Game, University of Lancaster, pers. comm. 1978

/ drawings overleaf

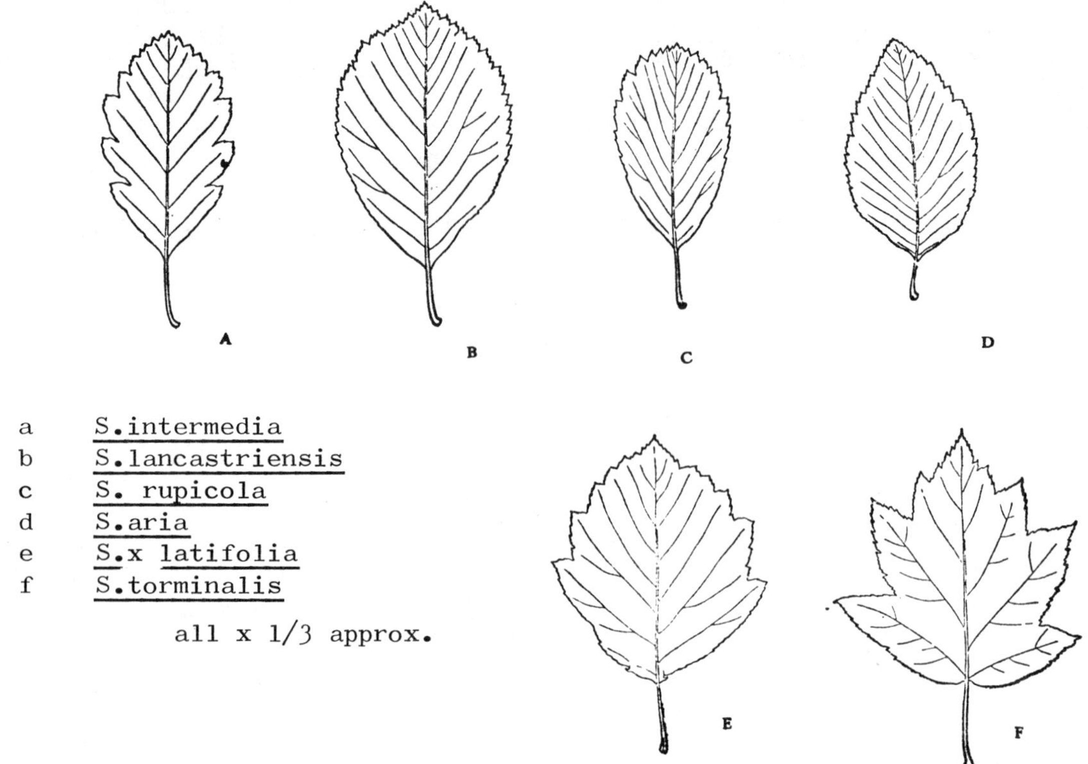


a S.intermedia
b S.lancastriensis
c S. rupicola
d S.aria
e S.x latifolia
f S.torminalis

all x 1/3 approx.

MALUS SYLVESTRIS Mill.

The two subspecies can be separated as follows:

Leaves with a few scattered hairs beneath, becoming glabrous; pedicels, receptacle and outside of calyx glabrous or nearly so. subsp. sylvestris

Leaves persistently hairy beneath; pedicels, receptacle, and outside of calyx tomentose. subsp. mitis (Wallr.) Mansf. (M.domestica Borkh).

Most of the garden apples can be referred to subsp.mitis.

Intermediates in the degree of hairiness are widespread.

RIBES RUBRUM L. / R. SPICATUM Robson

Although the leaves of the two segregates differ, it is unsafe without much experience, to separate them on leaf-shape alone. Even then, the full spectrum of leaves on any plant should be taken into account.

Inflorescence drooping; receptacle obscurely pentagonal with a raised rim around the style - in section shallowly-concave; anther connective-tissue broad. R. rubrum L. (R. sylvestris (Lam.)Mert. & Koch)

Inflorescence upright at first, spreading or arching in flower; receptacle perfectly circular, without a raised rim - in section deeply-concave; anther connective-tissue very narrow on inner side, broader on outer side. R. spicatum Robson

The drawings in Ross-Craig are good, except that the receptacle of R. rubrum is not always as clearly pentagonal as shown (Fig.D, below). The anther connective character is diagnostic in fresh material.

R. spicatum is a scarce calcicole of woodlands and limestone pavement.

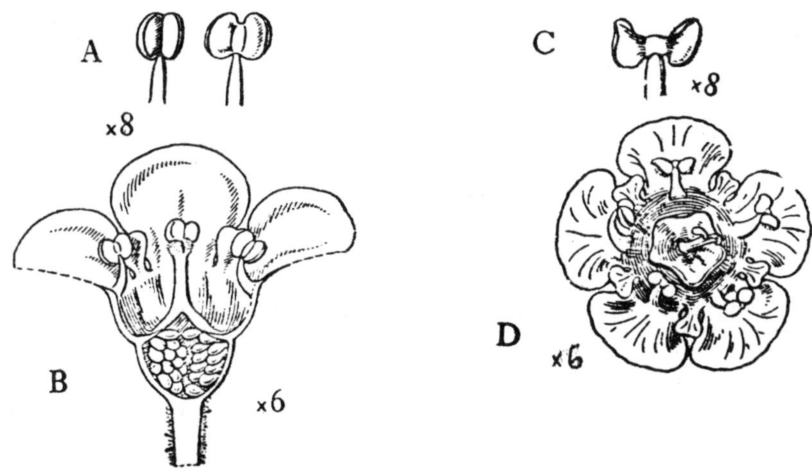


A,B: anthers and flower of R. spicatum
C,D: anther and flower of R. rubrum

EPILOBIUM L.

Difficulties in identification of the species arise mainly through the great phenotypic plasticity of many characters, particularly the leaves and the quantity of the indumentum, and the failure of some botanists to examine critically proven diagnostic characters, notably the quality of the indumentum.

Hybridisation is of widespread, though rather sporadic, occurrence in this genus. Hybrids are intermediate between the parents, but the extent of intermediacy is unpredictably variable. Ash (1953) listed some good hybrid indicators: taller & more branched habit, flowers unusually large or small in size, petals markedly deeper in colour at their tips, fruit shortened and undeveloped with mostly abortive seeds. Hybrids are mostly found in disturbed habitats such as quarries, waste land, neglected gardens and allotments. All plants that cannot be clearly assigned to a particular species should be collected for determination.

Particular care is needed in the E. tetragonum agg. The glandular hairs of E. obscurum may be very few and/or confined to grooves along the calyx tube, and therefore not showing up in silhouette.

1. Stem and inflorescence with crisped hairs, and numerous short, spreading, glandular hairs; flowers small, 4-6mm in diameter; stigma less than half the length of the style. E.ciliatum Raf. (E. adenocaulon Hausskn.)

1. Stem and inflorescence without glandular hairs, or only on the calyx-tube (sometimes very few); flowers 6-9mm in diameter; stigma about equalling the style.

2. Plant wholly without glandular hairs; capsule 5-8cm or more; plants forming almost sessile leaf rosettes at the base of the stem in autumn. Plants not producing stolons. E. tetragonum L. agg. (E.adnatum Griseb.)

3. Leaves oblong, ± decurrent. E.tetragonum L. subsp.tetragonum

3. Leaves mostly oblong-lanceolate, shortly petiolate, not decurrent. E.tetragonum L. subsp.lamyi (F.W.Schultz)Nyman

2. Calyx-tube with glandular hairs at least in the furrows sometimes very few; capsule 4-6cm long; plants producing stolons in summer. E. obscurum Schreb.

Stace, 1975.
G.M. Ash, in J.E. Lousley, The Changing Flora of Britain.

## EPILOBIUM ALSINIFOLIUM Vill. / E. ANAGALLIDIFOLIUM Lam.

The two alpine Epilobium species occur in our region, and though generally distinct, can be alike. E. alsinifolium is usually the larger plant, but may be as small as E. anagallidifolium.

| | E. alsinifolium | E. anagallidifolium |
|---|---|---|
| plants | often more upright, i.e., decumbent, then strongly ascending | usually more creeping, or more gradually ascending |
| stems | 5-20(30)cm long, 2-3mm in diam. | 4-10(20)cm long, 1-2mm in diam. |
| stolons | below ground, yellowish, with distant pairs of yellowish scales. | above ground, with distant pairs of small green leaves |
| leaves | usually 1.5-4.0 x 1.0-2.5 cm, somewhat bluish-green, distantly sinuate-toothed | usually 1.0-2.5 x 0.5-0.7cm, often yellowish-green, entire, or faintly sinuate-toothed |
| flowers | 8-9mm in diameter | 4-5mm in diameter |
| seeds | 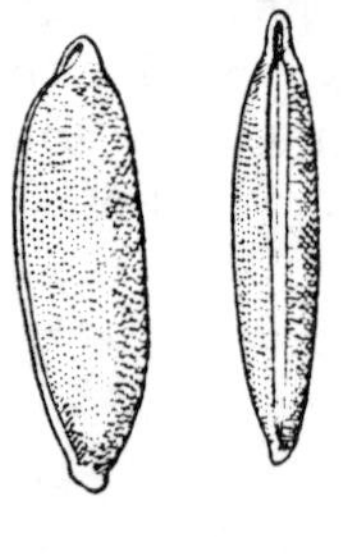 x16 | 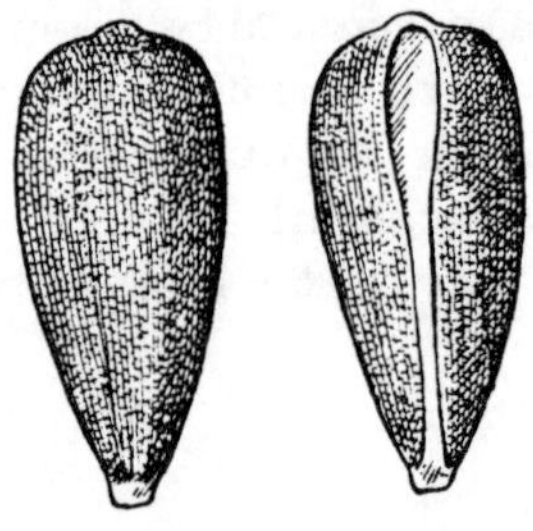 x16 |

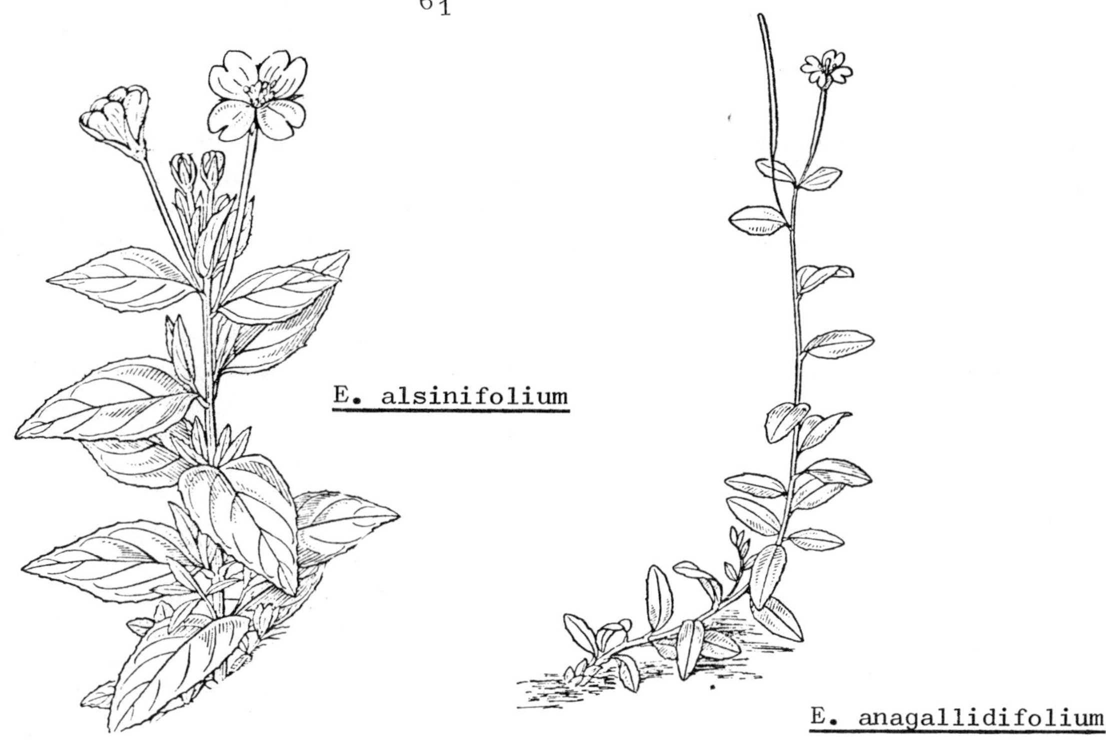


Hybrids recorded in the region:

E. x intermedium Ruhmer (E.hirsutum x E.parviflorum)
E. x erroneum Hausskn. (E.hirsutum x E.montanum)
E. x limosum Schur. (E.montanum x E.parviflorum)
E. x dacicum Borbas (E.obscurum x E.parviflorum)
E. x montaniforme Knaf ex Célak (E.montanum x E.palustre)
E. x rivulare Wahlenb. (E.palustre x E.parviflorum)
E. x haynaldianum Hausskn. (E.alsinifolium x E.palustre)

EPILOBIUM PEDUNCULARE agg.

Leaves with veins showing faintly beneath; seeds papillose, 0.75-0.90mm. Widespread, and often common in wet places in upland areas.
E.brunnescens (Cockayne) Raven & Englehorn subsp. brunnescens (E.nerteroides A.Cunn.) (E.pedunculare auct.)

Leaves with prominent veins showing above, and not showing beneath; seeds smooth, 0.5-0.9 mm. Recorded in dockland in v.c.66. E.komarovianum Léveillé

OENOTHERA L.

The taxonomy of Oenothera has been a matter of some dispute in the past, but recent work of K. Rostański has brought some order to the genus. Species of Oenothera are found in waste places, roadsides, railway banks, dunes etc., and it seems likely we have few species in our region (perhaps only O.biennis, O. erythrosepala, and their hybrid).

Stem without red-based bulbous hairs; basal leaves narrowly-oblanceolate; sepals green; petals 18-25mm; capsule pubescent, lacking red-based bulbous hairs. O. biennis L.

Stem with red-based bulbous hairs; basal leaves broadly elliptic-oblanceolate; sepals red-striped; petals 40-50mm; capsule pubescent, with red-based bulbous hairs. O. erythrosepala Borbas

O. x fallax Renner em. Rost. (O. biennis x O. erythrosepala) can be recognised by having the smaller flowers of O. biennis, but the red-striped sepals of O. erythrosepala, and at least a few red-based bulbous hairs. Hybrids are apparently fertile, and can give rise to swarms showing all degrees of variation between the parents. It has not been seen on our region, though recorded in West Lancashire.

Short notes, Watsonia, 12(2), 164-165, (1978)
C.A. Stace (1975)

CIRCAEA L.

Some of the characters hitherto used to distinguish the two species of Circaea are of little or no use. Among them may be mentioned the winged petioles supposed to be found in C. alpina, and the distinctions in the stigmas described for these species. The hybrid is known by its intermediacy and sterility.

1. Inflorescence not elongating until the petals have dropped, the open flowers clustered at the apex; fruit 2 x 1mm, uniocular; plant 5-20cm high; leaves cordate at base, dentate. C. alpina L.

1. Inflorescence elongating before the petals have dropped, the open flowers well-spaced; fruit ± bilocular; plant 10-60cm high; leaves truncate to shallowly-cordate at base, sparsely denticulate to dentate.

2. Plants fertile; fruit 3-4 x 2.0-2.5mm; plant 15-60cm high; leaves truncate to slightly cordate at base, sparsely denticulate. C. lutetiana L.

2. Plants sterile; fruit to 2.0 x 1.2mm, falling in the immature state, with one loculus abortive; plant 10-45cm high; leaves shallowly-cordate at base, dentate. C. x intermedia Ehrh.
(C.alpina x C.lutetiana)

P.H. Raven, in Flora Europaea, Vol.2, (1968)

CALLITRICHE L.

The identification of Callitriche species often presents the recorder with considerable difficulties, for three main reasons:

i) the frequent absence or scarcity of fruiting plants
ii) the variability of leaf characters (whose morphology depends greatly upon the habitat
iii) the requirement for microscopical examination

Although leaves are very variable, plants can often be assigned to a particular species on leaf-shape alone. In many cases, however it is necessary to confirm the diagnosis by reference to fruiting and other characters. Terrestrial forms usually differ greatly in morphology from the aquatic forms, and even when fruit is present, such plants can be impossible to name with certainty.

The anatomical details of the mericarps (= seeds in CTW and Ross-Craig) can be observed by stripping off the outer layer of cells.

1. Terrestrial.

2. Mericarp with rounded margin, not winged or keeled; pollen grains oblong-ellipsoid or slightly reniform (more than 70% at least twice as long as broad). C. obtusangula LeGall

2. Mericarp winged or at least bluntly-keeled; pollen grains variously-shaped, but few or none twice as long as broad.

3. Styles persistent, reflexed and closely appressed to sides of fruit; stamens short, 1-2mm. C. hamulata Kutz. ex Koch (C. intermedia Hoffm.)

3. Styles persistent, erect or recurved, not appressed to sides of fruit; stamens 2-3mm or more. It is not always possible to proceed further - see note below.

4. Styles arcuate-recurved; fruit pale brownish; mericarps broadly-winged; pollen-grains sub-globose, 18-24μm, all viable. C. stagnalis Scop.

4. Styles erect or patent; fruit brown; mericarps narrowly-winged; pollen-grains variously-shaped, 24-30μm, usually 15-30% sterile. C. platycarpa Kutz. (C. verna, auct.)

1. Aquatic.

5. All leaves submerged.

6. Leaves emarginate at apex; peltate hairs and stomata absent; styles caducous, recurved, not closely-appressed to sides of fruit; mericarps broadly-winged. C. hermaphroditica L.

6. Leaves with expanded, deeply emarginate ('spanner-like') apex; peltate hairs with 10-15(18) radiate cells; stomata present; styles persistent, reflexed and closely appressed to sides of fruit; mericarps narrowly-winged. C. hamulata Kutz ex Koch (C. intermedia Hoffm.)

5. Upper leaves forming a floating rosette.

7. Submerged leaves linear, with expanded, deeply-emarginate ('spanner-like') apex; peltate hairs with 10-15(18) radiate cells; styles reflexed and closely appressed to sides of fruit; mericarps narrowly-winged; stamens short, 1-2mm; anthers up to 2mm; pollen-grains colourless. C. hamulata Kutz ex Koch (C. intermedia Hoffm.)

7. Submerged leaves linear, narrowly-elliptical, or narrowly-rhombic, not with 'spanner-like' apices; peltate hairs with (6)8-10(12) radiate cells; styles recurved or patent, not closely appressed to sides of fruit; mericarps broadly- or narrowly-winged; stamens 2-3mm or more; anthers usually more than 2mm; pollen grains yellow.

8. Rosette-leaves rhombic, distinctly 3-ridged, giving the rosette a distinctive, corrugated appearance; submerged leaves narrowly-rhomboidal, becoming linear below; fruit-lobes with rounded margins; mericarps with rounded, scarcely discernible margin; pollen grains oblong-ellipsoid, more than 70% at least twice as long as broad. C. obtusangula LeGall

8. Rosette-leaves not rhombic; submerged leaves narrowly-elliptical or linear; fruit-lobes keeled; mericarps winged; pollen grains never twice as long as broad.

9. Leaves pale-green, rosette-leaves broadly-elliptical or suborbicular, about twice as long as wide, submerged leaves narrowly-elliptical, never linear; fruit pale brownish; mericarps broadly-winged; stamens 2-3mm; pollen grains subglobose, 18-24μm, all viable. C. stagnalis Scop.

9. Leaves often a deeper green (often with a bluish tinge), rosette-leaves elliptical, 2-4 x as long as wide, some or all submerged leaves linear, with emarginate apex; fruit brown; mericarps keeled or narrowly-winged; stamens 3mm; pollen grains variously-shaped, 24-30μm, usually 15-30% sterile. C. platycarpa Kutz (C. verna, auct.)

The most difficult separation is that of C. platycarpa and C. stagnalis. Lewis-Jones & Kay found in Glamorgan that they could not unequivocably separate them on the basis of fruit morphology alone, but found that a chromosome count was the most reliable distinction.

However, they can be separated on leaf characters when typical, Pollen grain-size, shape, and viability appear to be useful differentiating characters.

C. platycarpa is a widespread species and probably more so than the records suggest. It is mainly confined to lowland areas below 250m.

C. brutia Petagna (C. intermedia subsp. pedunculata (DC.) Clapham) is sometimes given specific status. It is very similar to C. hamulata, but is usually smaller, and is usually found in shallower water or on mud. The submerged leaves are less expanded and the apex is often unequally bifid; the fruit-stalks are up to 13mm long. The distribution of the taxa is imperfectly known, and it may occur in the region.

H.D. Schotsman, in Flora Europaea, Vol.3 (1972)
L.J. Lewis-Jones & Q.O.N. Kay, Nature in Wales 15, 180-183 (1977)
J.P. Savidge, pers. comm. (1974)

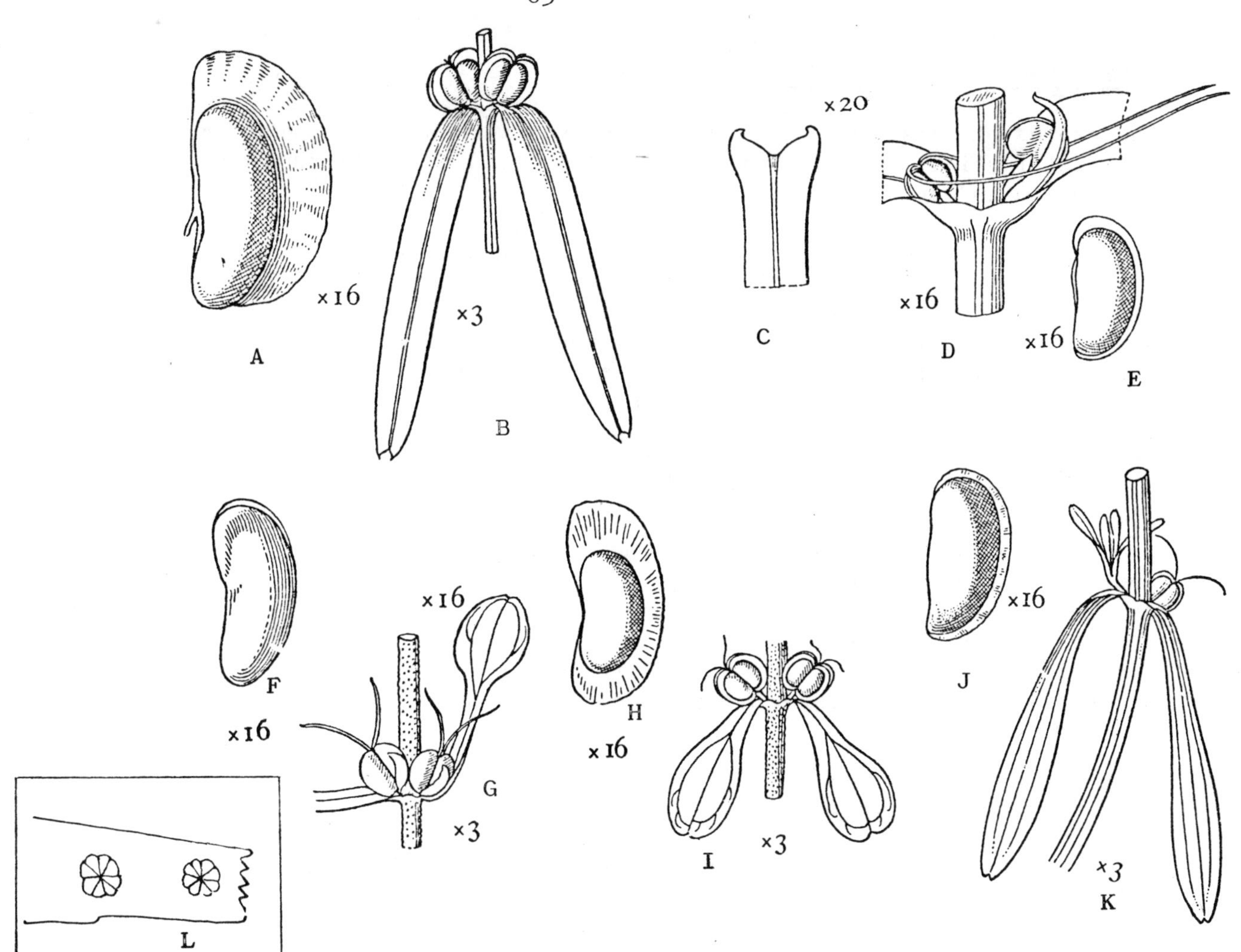


C. hermaphroditica (A,B) C. hamulata (C,D,E)
C. obtusangula (F,G) C. stagnalis (H,I)
C. platycarpa (J,K) Peltate hairs x200 (L) on part of leaf

MYRIOPHYLLUM L.

Three species of Myriophyllum occur in our area. Two are frequent: M. spicatum especially in the lowlands, and M. alterniflorum particularly in upland areas. There are several old records for M. verticillatum in Cumbria.

1. Leaves 25-45mm long, usually 5 in a whorl, often longer than the internodes, segments 24-35; turions (winter-buds) formed in autumn; bracts pinnatisect, as long as the flowers even near the top of the spike. M. verticillatum L.

1. Leaves usually 4 in a whorl, about equalling the internodes, segments 6-38; turions 0; upper bracts simple, entire or serrate.

2. Leaves 15-30cm long, with 6-18 segments; spike short (not more than 3cm), at first drooping at the tip; flowers whorled at the base, solitary or in opposite pairs above. A plant of base-poor waters. M. alterniflorum DC.

2. Leaves 10-25mm long, with 13-35 segments; spike usually more than 4cm long, erect throughout; all flowers whorled. A plant of base-enriched waters. M. spicatum L.

<u>HEDERA HELIX</u> L. var. <u>HIBERNICA</u> Kirchin

The status of this variety is uncertain, and work has been done recently on its distribution. It is common in Scotland, and may well turn out to be widespread in our region.

The Irish Ivy is distinguished from the common variety by its larger, usually 5-lobed leaves which are wider than long (5-15 x 8-10cm ). The leaves are pliant, waxy and of a softer texture, and a pale, richer green. The lobes are broadly triangular, the angles between them wide, and the leaf-bases deeply cordate. The petioles are proportionately longer (up to 20cm ), green or tinged pink-buff. Inflorescence slightly larger and more pubescent. It is a very vigorous grower, but often by-passes trees in its path. It forms dense sheets, with leaves of uniform size and shape, or when growing from a wall, cascades luxuriantly in lax curtains quite unattached except at the roots. As the leaves are not quite flat, the light strikes part of the leaf only, making the plant eye-catching even from a distance. It has considerably less anthocyanin, and scarcely bronzes in cold weather.

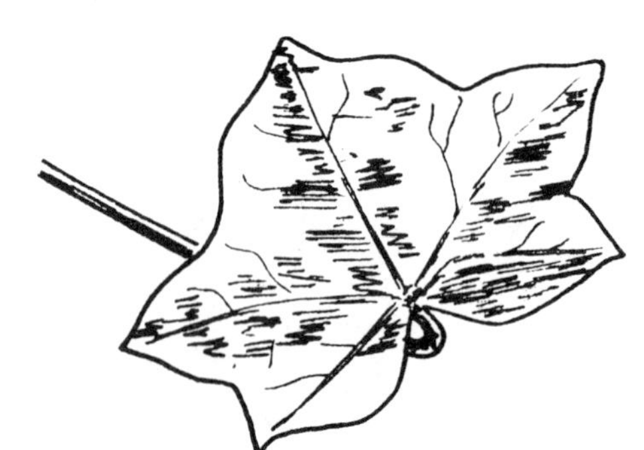

<u>BUPLEURUM ROTUNDIFOLIUM</u> L./<u>B. SUBOVATUM</u> Link. & Spreng.

<u>B. subovatum</u> Link. & Spreng. is occasionally found in the region, growing in waste places, gardens etc. It is commonly mis-identified as <u>B. rotundifolium</u> L. which is a rare native plant of cornfields, with no recent records in our region. Both species have perfoliate leaves, and are without bracts.

Rays (3) 5-10; bracteoles oblanceolate to ovate; fruit smooth.
<u>B. rotundifolium</u> L.

Rays 2-3(5); bracteoles suborbicular; fruit conspicuously tuberculate; leaves usually relatively narrower. <u>B. subovatum</u> Link. & Spreng.

In the past, <u>B. subovatum</u> has been erroneously named <u>B. lancifolium</u> Hornem. These names are not synonymous and <u>B. lancifolium</u> is not known in Britain.

<u>APIUM NODIFLORUM</u> (L.) Lag./<u>BERULA ERECTA</u> (Huds.)Coville

Care should be taken in distinguishing these two species vegetatively. Leaf-shape, serration, and number of leaflets is variable in both species, as also are height and habit. The presence or absence of a septum on the petiole is diagnostic.

Petiole lacking discoloured 'ring-mark' or septum, and rudimentary leaflets.
<u>Apium nodiflorum</u> (L.) Lag.

Petiole with discoloured 'ring-mark' or septum some distance below the lowest pair of leaflets which are sometimes rudimentary.
<u>Berula erecta</u> (Huds.) Coville

When flowering, the two species are easily distinguished ; <u>Berula erecta</u> has conspicuous bracteoles, and <u>Apium nodiflorum</u> has none.

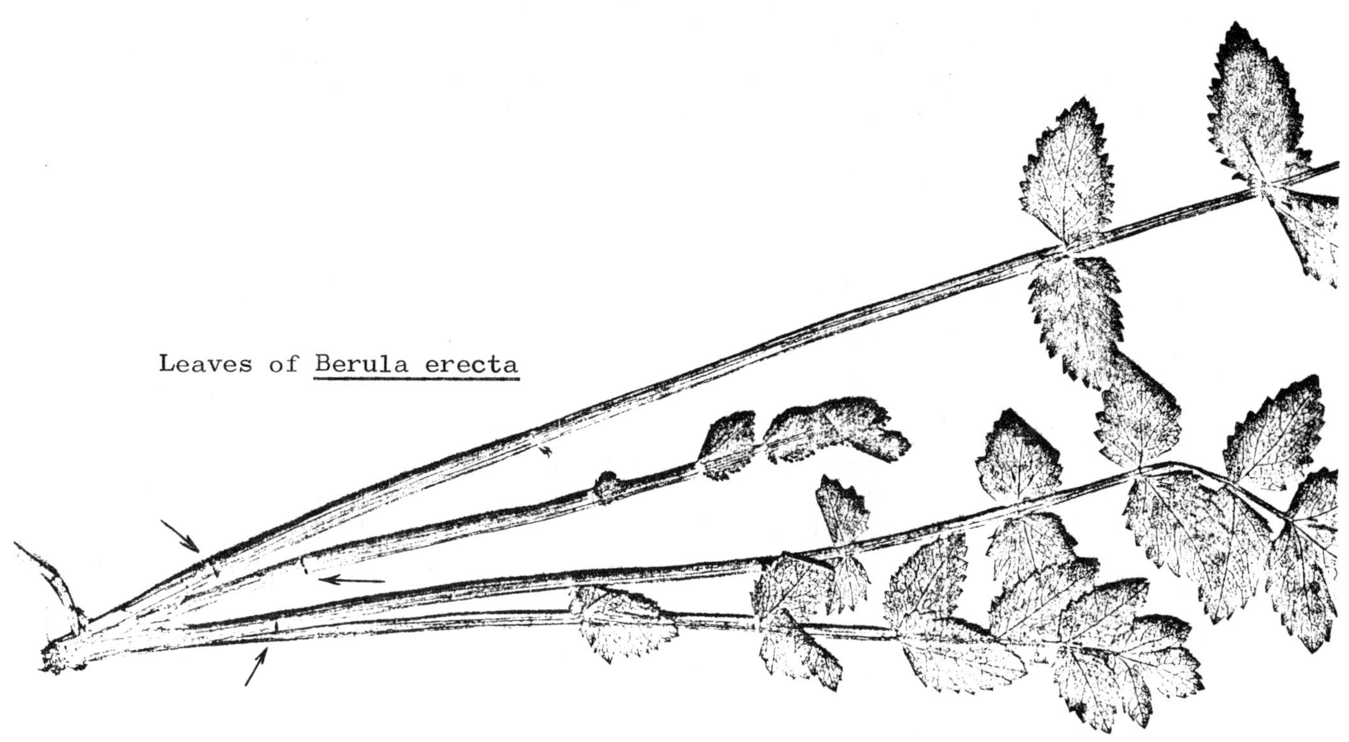


HERACLEUM MANTEGAZZIANUM Somm. & Levier x H. SPHONDYLIUM L.

The hybrid is intermediate in size of stem, leaf, and umbel, in leaf-outline, in the shape of fruit and vittae, in the hairiness of the stem and sheath, and in the smell when bruised. Hybrids have reduced or no fertility.

Most records of the hybrid are from Scotland, but it is probably under-recorded and may occur in our region.

POLYGONUM AVICULARE L.

Characters of the fruit and the persistent perianth are the most reliable differentiating features. There is, however, wide variation even on a single plant. Maritime forms of P. aviculare in this condition have frequently been erroneously named P. raii.

1. Fruit distinctly shining, smooth, 5.0-6.0 x 3.0-3.5mm, longer than the persistent perianth; ochreae with 4-6 unbranched veins, shorter than internodes. Coastal - on fine shingle or sandy shores. P. oxyspermum Meyer & Bunge ex Ledeb. subsp. raii (Bab.) Webb & Chater (P. raii Bab.)

1. Fruit dull (or shiny on the margins), striate, enclosed by, or slightly longer than the persistent perianth.

2. Branch leaves much smaller than the stem leaves; persistent perianth divided almost to base; fruit trigonous with 3 concave sides.

3. Stem leaves narrow, linear-lanceolate, 1-4mm broad; perianth segments narrow; fruit 2.5-3.5 x 1.5-2.0mm, narrow and exserted. In calcareous arable fields. P. rurivagum Jord. ex Bor.

3. More upright in habit than P. arenastrum (which is prostrate and mat-forming). Stem leaves ovate-lanceolate, sub-sessile; leaves 20-50 x 5-15mm, those of the main stem 2-3 times as long as those on the flowering branches; petioles about 2mm, included in the ochreae; perianth divided for more than half its length; fruit 2.5-3.5 x 1.5mm, punctate, dull-brown, with 3 equal sides. P. aviculave L.

2. Branch and stem leaves all ± equal; persistent perianth divided for half its length; fruit 1.5-2.5mm, with two sides convex and slightly larger than the one concave face; leaves up to 20 x 5mm.
P. arenastrum Bor.

P. arenastrum is probably under-recorded in our region, though doubtless wide-spread and often locally common. P. oxyspermum subsp. raii is found in Cumbria and Northumbria, formally also in Durham. P. rurivagum may perhaps still occur; there is one old record for v.c.66.

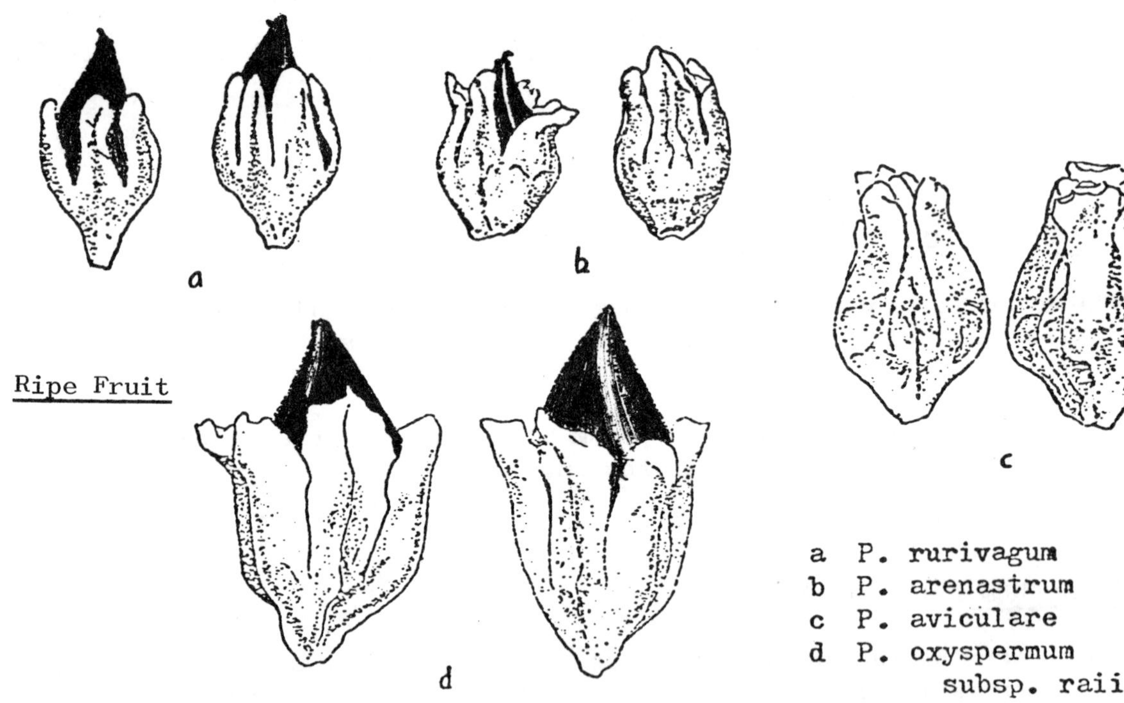


a P. rurivagum
b P. arenastrum
c P. aviculare
d P. oxyspermum subsp. raii

mainly B.T. Styles, Watsonia 5(4) 177-214 (1962)

RUMEX ACETOSELLA L. agg.

The two segregates are separated as follows:

Stems erect; lower leaves 3-4 times as long as broad (excluding lobes), margins flat; ripe fruit about 1.5 x 0.8mm. Common, avoiding the poorest soils, and less strongly calcifuge than R. tenuifolius. R. acetosella L.

Stems decumbent at the base; lower leaves 7-10 times as long as broad (excluding lobes), margins inrolled; ripe fruit about 1.0 x 0.7mm. Uncommon, on the poorest acidic soils. R. tenuifolius (Wallr.) Löve

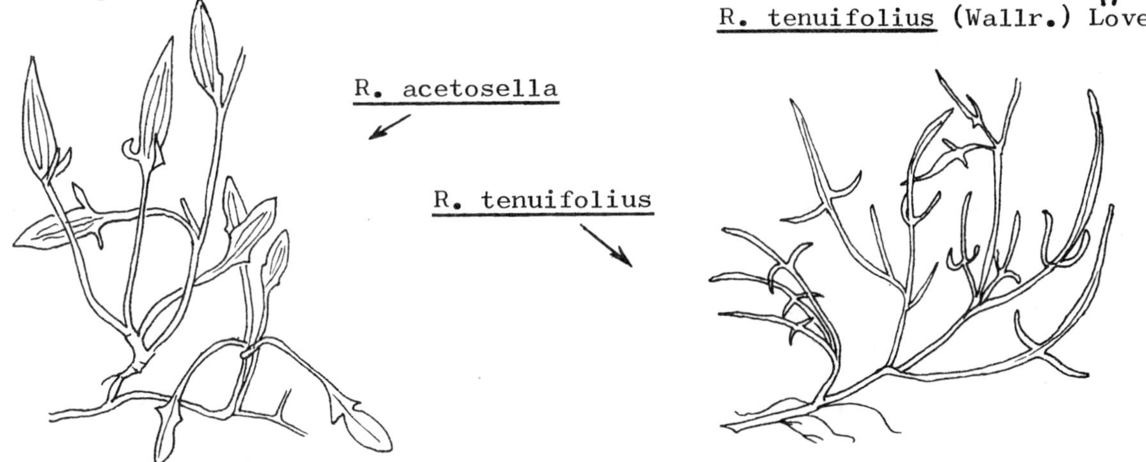


RUMEX L.

Hybrids in the subgenus Rumex are frequent. They are usually recognised by the irregular enlargement of the tepals, of which only a small proportion mature. The remainder drop off when only partially developed. The panicle is often flushed with red. Partial sterility shows also in the pollen with a high proportion of shrivelled grains. The flowering period of hybrids is often extended into the autumn.

Rumex hybrids have probably been much overlooked in our region, though some may be frequent. Those plants appearing intermediate and/or with the above general characters should be collected for determination.

R. x pratensis Mert. & Koch (R. crispus L. x R. obtusifolius L.) is probably common in our region where the parent species grow together, but it is so far recorded only in Durham. It is intermediate between the parents in:

i) basal leaf length/breadth ratio, hairiness, and number of main lateral veins

ii) cauline leaf hairiness, and number of veins

iii) shape of stipules and tepals

iv) number of tubercles

R. x arnottii Druce (R. longifolius DC. x R. obtusifolius L.) is known in Upper Teesdale, Durham.

Hybrids between these and other common species should be looked for, and collected for determination.

J.E. Lousley and J.E. Williams, in Stace (1975)

BETULA L.

The taxonomy of the genus Betula is disputed, and opinions differ as to the extent of hybridisation. Intermediates between B. pendula and B. pubescens are, however, common and widespread. They are intermediate in leaf-shape, pubescence, bark of trunk and twigs, habit, and fruiting catkins. Most intermediates seem fully fertile and probably backcross, since the intermediates are not always clearly separable from the parent species.

1. Tree or large shrub, usually more than 3m high. Young male catkins in winter pendent and unprotected.

2. Bark at first shiny and red-brown, later pinkish-white with horizontal, broad, pale-grey bands and some dark-grey scaling patches, finally white with large, black diamonds, often deeply-fissured at the base into small, black, knobbly plates. Young twigs glabrous, slender and pendent, with numerous peltate, non-odorous resin glands or warts. Leaves usually sharply biserrate with prominent primary teeth, glabrous. Nutlet glabrous. B. pendula Roth. (B. verrucosa Ehrh.)

2. Bark at first red-brown as in B. pendula, becoming smooth, greyish-white, variably banded horizontally grey or brown (sometimes with fine, grey lacework patterns), often remaining brownish until mature, but never with black diamonds. Young twigs usually ± pubescent, spreading or ascending, with or without resinous glands. Leaves irregularly-serrate, without prominent primary teeth, usually hairy, at least on the veins beneath. Nutlet puberulent at apex.

3. Tree with a single stem. Young twigs usually conspicuously pubescent. Buds not viscid. Leaves 3-4cm. B. pubescens Ehrh. subsp. pubescens

3. Shrub or small tree, often with several stems. Young twigs glabrescent, covered with brown, sticky, resinous, pleasantly-smelling glands. Buds viscid. Leaves often less than 3cm. long. Most readily distinguished in spring when the unfolding buds have a pleasant, resinous smell. B. pubescens subsp. carpatica (B. pubescens subsp. odorata (Bechst.) E.F. Warb.)

1. Small shrub, rarely more than 1m. Young male catkins in winter, erect, protected by bud-scales. Leaves orbicular, obtuse or truncate. Petiole 3mm or less. One locality in the region. B. nana L.

## ALNUS Mill.

*Alnus incana* is sometimes planted in the region for shelter, or for the re-colonisation of pit-heaps. From *A. glutinosa* it is easily distinguished by its acute leaves.

Leaves truncate or retuse at apex, bright green on both sides, with 4-7 pairs of veins; bark dark-brown. *A. glutinosa* (L.) Gaertn.

Leaves acute or acuminate, pale or glaucous beneath, with 10-15 pairs of veins; bark pale grey. *A. incana* (L.) Moench.

*A. glutinosa*

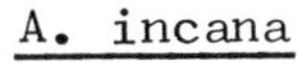

*A. incana*

## QUERCUS L.

There is considerable hybridisation and introgression between our two native species of oak. Trees with intermediate characters are widespread and common in our region, and are variable and fertile. The limits of the true species, and the extent of mixing are still a matter of opinion. However, one can come to recognise intermediates and the true species (or near) on a *combination* of leaf characters.

The tabular key may be used equally for fresh material or for leaf litter, and thus oaks may be determined throughout the year. The number of lobe pairs on a leaf is counted from the base of the lamina to the tip. In practice, all lobes are counted, the number divided by two, and any halves discounted. i.e., $4\frac{1}{2}$ lobe pairs count as 4 pairs.

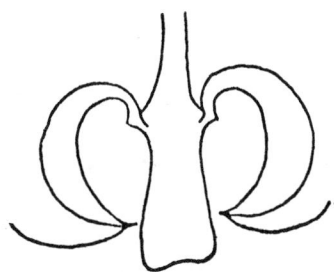

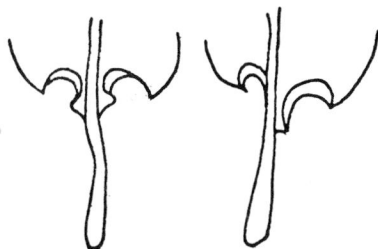

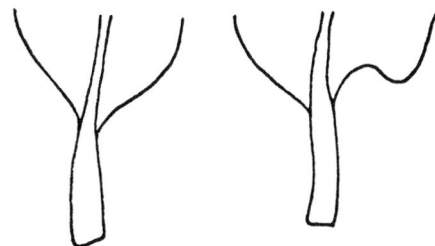

*Q. robur* Intermediates *Q. petraea*

*Leaf-bases*

| | Q. robur type | Intermediates | Q. petraea type |
|---|---|---|---|
| lobing | deep, irregular | shallow, irregular, or deep, regular | shallow, regular |
| lobe pair number | 4 or less | 5 | 6 or more |
| leaf base | strong auricles | intermediate auricles | weak auricles, or none |
| abaxial pubescence | no stellate hairs present | a few stellate hairs may be present | large and small stellate hairs present |
| petiole percentage | 2-8% leaf length | intermediate length | 10-20% leaf length |
| leaf shape | oblanceolate | variable | elliptic, or nearly so |

D.L. Wigston, Watsonia 10 345-369 (1975)

POPULUS L.

The genus Populus presents particular problems of identification, and these have been considerably increased by the introduction of many wild and horticultural varieties. As a result, this genus (with the exception of P. alba, P. x canescens, and P. tremula) has been neglected, and little is known of the occurrence and distribution of the other taxa in our region. The situation is further complicated by the presence of an involved synonymy, and the names used here are those generally adopted in the more recent texts.

The native black (or Manchester) poplar (P. nigra var. betulifolia) does not occur as a native tree in our region, but long-established trees indistinguishable from the native stock, are occasionally seen in avenues or as single planted trees.

Hybrid black poplars (P. x euramericana, P. x canadensis) derive from the European black poplar, P. nigra, and the N. American P. deltoidea. They are more vigorous than either parent, and are widely planted. They are sometimes mistakenly recorded (because of the English name) as P. nigra.

1. Tree narrowly-columnar; branching fastigiate (i.e. ± erect); bark dull-grey, shallowly-ridged. P. nigra 'Italica'

1. Tree domed or widely-spreading, not narrowly-columnar; not fastigiately branched.

2. Bark on branches partly cream or white.

3. Lightly-branched, leaning, rather flat-topped; leaves silvery below, five-lobed on strong growth. P. alba L.

3. Massively-branched, tall, high-domed; shoots at first covered in scurfy white pubescence; leaves orbicular or nearly so with large, irregular, curved, blunt teeth. P. x canescens (Ait.) Sm. (P. alba x P. tremula)

2. Bark uniformly grey or brown.

4. Upper branches short, horizontal; crown narrowing upwards; leaves ovate-orbicular, fluttering to show light greenish-grey undersides.
P. tremula L.

4. Upper branches ascending, forming a large, broad dome, or flat top.

5. Bark fissured, or if fairly smooth, then buds not viscid, and with no balsamic fragrance.

6. Main branches arch outwards.

7. Trunk develops large bosses or burrs; bark brown; branches massive; with upswept, dense bunches of straight shoots from the upperside of the deeply-arched, widely-spreading branches.
P. nigra L. var. betulifolia (Pursh.) Torrey

7. Tree relatively narrow, open-topped; bark grey; branches not massive; branches arch outwards at the ends to form a vase shaped crown; shoots slender, partly pendulous; with many snags and dead shoots; leaves pure green. Female trees only.
P. x euramericana (Dode) Guinier
'Regenerata' (the railway poplar)

6. Main branches ascending or nearly vertical; large tree with open, rather gaunt habit, and extremely vigorous; bole and branches clean; branches curving upwards to form a goblet-shaped crown; leaves grey-green. Male only. The black Italian poplar.
P. x euramericana 'Serotina'
(P. x canadensis Moench)

5. Bark smooth, or shallowly-ridged; buds viscid, and with balsamic fragrance; leaves whitish below. (the balsam poplars).

8. Leaves broad, strongly cordate, strongly pubescent on petiole and on the veins beneath; branches more spreading, producing a broader, more open crown. Female clone only.
P. candicans Ait.
(P. x gileadensis Rouleau)

8. Leaves neither strongly cordate at base, nor strongly pubescent; leaves sometimes very large - to 35 x 25cm.

9. Numerous slender suckers around the base for many yards; bark with pink fissures, cleaner bole; branches more upswept; shoots not winged or angled.
P. balsamifera L.
(P. tacamahaca Mill.)

9. Suckers absent; bark smooth until tree quite large; young trees narrowly conic, older trees with broad, brush-like tops, as branches nearly upright; most shoots angled.
P. trichocarpa Hook.

The balsam poplars

Several species of balsam poplar are susceptible to bacterial canker, and have become rare. True species occur uncommonly in the region, and have been largely replaced by hybrids. These are generally intermediate in leaf-shape, pubescence (if any), petiole length, habit, etc. The hybrid most commonly planted is 'TT32' (P. trichocarpa x P. balsamifera).

A. Mitchell (1974)
W.J. Bean. Trees and shrubs hardy in the British Isles. 8th edn. John Murray 1976.

SALIX L.

The problems of identification of the taxa in this complex genus are considerably increased by the planting of horticultural cultivars. Dihybrids and trihybrids of the native species occur, and the hybrids which have already been recorded in the region are listed in the appendix.

Anyone with a good eye for the true species may detect hybrids with more ease than is the case in many other critical genera. The correct determination of these hybrids is, of course, another matter, and good material of suspected hybrids should be collected for expert determination. A spray with undamaged leaves is the minimum requirement. The same plant should preferably be sampled twice, in catkin, and in full leaf.

Any relevant notes should be added, both about the willow sampled, and of other willows growing nearby.

The descriptions of the hybrids occurring in our region cannot be given here, but those interested should refer to R.D. Meikle, in Stace, 1975. All hybrids recorded in the region are listed in the Appendix.

VACCINIUM x INTERMEDIUM Ruthe (V. myrtillus L. x V. vitis-idaea L.)

This hybrid has not yet been found north of the southern Pennines, but it would be worthwhile looking for it in moorland areas in our region. It is intermediate between the parents in many features, including;

i) degree of ridging of the stem - i.e. slightly ridged or angled
ii) colour, shape, thickness, toothing and persistence of leaves
iii) number, shape, colour of flowers - (1-)2-3(-4), corolla lobes $<\frac{1}{2}$ corolla
iv) colour of fruit - reddish-black or purple - rare.

It shows some hybrid vigour, some patches spreading at the expense of the parent species. Its habitats seem almost always to have been recently disturbed by man - banks of cut peat, edges of ditches or drains, old cart-tracks or moorland paths, old gun-sites etc.

PYROLA MINOR L./P. MEDIA Sw.

Pyrola minor and P. media can be very similar, with several characters less well differentiated than is suggested by CTW (such as leaf-size, flower size and colour).

The best distinguishing features are, style length and shape, and anther length.

Style 1-2mm, not expanded into a disc below the stigma, included; anthers 0.8-1.2mm long. P. minor L.

Style 4-6mm, expanded into a disc below the stigma, exserted; anthers about 2.5mm long. P. media Sw.

There are no authenticated recent records of P. media in our region except in v.c.67.

D.A. Webb, in Flora Europaea, Vol.3 (1972)

PRIMULA x TOMMASINII Gren. & Godr. (P. vulgaris Huds. x P. veris L.)
(P. x variabilis Goupil, non Bast.)

This hybrid is sometimes found where the two parent species occur together. It is usually intermediate between the parents, but can be variable.

It differs from P. veris in having the leaves not abruptly contracted at the base; larger, paler yellow flowers with a less concave limb; and longer, more shaggy pubescence.

It differs from P. vulgaris in having a distinct scape and shorter pedicels; smaller deeper-yellow flowers with a more concave limb; shorter less shaggy pubescence.

P. vulgaris var. caulescens has a raised umbel on a scape, instead of being sessile and radical. It differs from the hybrid P. x tommasinii by its larger flowers, and its long, acute calyx teeth. There are no records of var. caulescens in our region.

LYSIMACHIA L.

All four of the erect Lysimachia species occur in the region. L.terrestris is known only from the Windermere area but may occur elsewhere. L.punctata and L.ciliata are more widespread, occurring occasionally in damp fields and marshy places.

1. Flowers terminal.

2. Flowers in a panicle; corolla without streaks or dots; axillary bulbils never present. L. vulgaris L.

2. Flowers in a raceme; corolla streaked and dotted with red and black; axillary bulbils usually present. L. terrestris (L.) Britton

1. Flowers axillary.

3. Flowers solitary, rarely paired; corolla with red basal blotches; leaf surfaces glabrous. L. ciliata L.

3. Flowers usually in clusters of 2 or more; corolla without blotches; leaf surfaces puberulent. L. punctata L.

L.F. Ferguson, in Flora Europaea, Vol.3 (1972)

ANAGALLIS L.

There are two blue pimpernels in Britain, one merely a colour form of the common A. arvensis. They can best be separated on the following characters:

Pedicels usually considerably exceeding the subtending leaf; calyx not concealing the corolla in bud; petals up to 6mm broad, fringed with 3-celled glandular hairs. A. arvensis L. f. azurea Hyl.

Pedicels not, or only slightly exceeding the subtending leaf; calyx concealing the corolla in bud; petals up to 3.5mm broad, fringed with 4-celled glandular hairs. Rare casual in the region. A. foemina Miller (A. arvensis L. subsp. caerulea Hartman)

Note that as f.azurea is merely a colour varient of A. arvensis, the normal varient (Scarlet Pimpernel) can always be used for reference in comparing the glandular hairs of the two taxa.

LIGUSTRUM L.

The introduced species, L. ovalifolium, is widely planted for hedging in the region, and isolated bushes are sometimes seen on waste ground etc. It is readily differentiated from the native species.

Young twigs puberulent; leaves 3-6 x 1-2cm, lanceolate, fairly thin, deciduous or semi-evergreen; panicles 3-6cm, puberulent; corolla tube equalling limb. Wood margins and scrub, mainly calcicolous. L. vulgare L.

Young twigs glabrous; leaves broader, usually more evergreen; panicles larger, glabrous; corolla tube 2-3 times as long as limb. Widely planted. L. ovalifolium Hassk.

CENTAURIUM ERYTHRAEA Rafn. / E. LITTORALE (D. Turner) Gilmour

The species are separated on many characters, including leaf- and stigma-shape, length of corolla-tube, and size of pollen-grains. Hybrids are intermediate between the parents in most characters, and have low pollen fertility (0-10%) and seed-set. Hybrids which appear to be backcrosses either to C. erythraea or to C. littorale usually show high pollen fertility (80-90%) and good seed-set.

| | C. erythraea | Intermediates | C. littorale |
|---|---|---|---|
| leaf length: breadth ratio | 2.1 - 3.3, short & broad | 3.4-4.9, inter-intermediate | 5.0-7.6, long & narrow |
| leaf-shape | elliptic, sides never parallel, 5-veined, apex acute | linear-elliptic, sides never parallel, apex acute, 3-veined | linear with sides parallel, 1-veined, apex obtuse |
| indumentum | glabrous | semi-scabrid | scabrid |
| calyx: corolla-tube ratio | 0.40 - 0.64 | 0.65 - 0.75 | 0.76 - 0.98 |
| corolla -tube length | 4.5 - 5.4mm | 5.5 - 5.6mm | 5.7 - 6.2mm |
| diameter of pollen-grains | 24 - 26mm | 27 - 28mm | 29 - 32mm |
| stigma-shape (see below) |  | | |

Hybrid plants are known in v.c. 59 and 60, and may, perhaps occur on the Cumbrian or Northumbrian coasts.

R. Ubsdell, Watsonia 11(1) 7-31 (1976)

CENTAURIUM ERYTHRAEA Rafn. /C. CAPITATUM (Willd.) Borbás

C. capitatum is known in the region only on the Northumberland coast, and is nationally rare. It can be very easily confused with dwarf plants of C. erythraea. The only sure distinction seems to be in the anthers, which are inserted at the base of the corolla-lobes in C. erythraea, and inserted at the base of the corolla-tube in C. capitatum. The corolla-tube must be split and carefully opened-out, to observe this character.

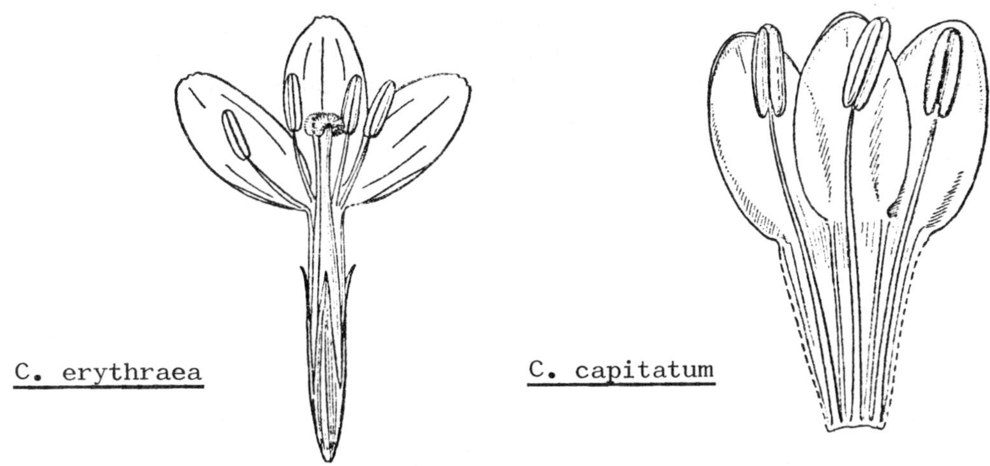

C. erythraea C. capitatum

SYMPHYTUM L.

The most frequent taxon in the region is the hybrid S. x uplandicum Nyman which forms a range of intermediates between the parent species. Specimens belonging to this group and not coming clearly into the spectrum of S. officinale should be placed in S. x uplandicum for the purposes of recording. Anything approaching S. asperum is unlikely to be found.

Stem with long, deflexed, conical hairs not swollen at base; leaves broadly-decurrent, middle and upper leaves sessile; corolla cream (var. ochroleuca) or reddish-purple (var. purpureum); anthers in front view longer than the exposed portion of the filaments. S. officinale L.

Stem with stout, hooked bristles very swollen at the base into small siliceous hemispheres; leaves not decurrent, petiolate or uppermost sessile; corolla sky-blue; anthers in front view shorter than the filaments.
S. asperum Lepech

The following account of S. x uplandicum is quoted from the Critical Supplement to the Atlas of the British Flora.

"This taxon is very variable in flower colour and leaf-decurrence, and probably arose as a result of hybridisation between S. officinale and S. asperum. Pure S. officinale has cream-coloured or reddish-purple flowers, and is more or less confined to fens and river banks. S. x uplandicum occurs in a wide range of habitats, particularly roadsides and hedgebanks, long stretches of which may be occupied by a single clone. This is particularly true of a form with clear blue flowers and non-decurrent leaves, which may possibly be an F1 hybrid.

In view of the rarity of S. asperum, which is always an introduction in Britain, it seems likely that S. x uplandicum has been introduced from Europe where it is grown as a fodder crop. It occurs naturally in southern Russia where the ranges of the parents overlap. The many variable forms of S. x uplandicum have possibly arisen as a result of backcrossing between the introduced S. x uplandicum and S. officinale".

F.H. Perring, Critical Supplement to the Atlas of the British Flora 1968, and pers. comm.

Three other species of Symphytum occur in the region.

S. orientale L. is softly-pubescent, has a white corolla, and calyx-teeth obtuse and only about ½ length of tube.

S. ibiricum Steven. (S. grandiflorum) has a yellow corolla, the calyx divided almost to base, and long slender rhizomes. It often forms small mats.

S. tuberosum L. has a yellowish-white corolla, the calyx-teeth about 3 x length of tube, and middle stem leaves much longer than the lowest. The rhizome is tuberous, not creeping, thus giving individual plants a very different appearance from the more herbaceous mats of S. ibiricum.

MYOSOTIS L.

Species of the water forget-me-nots are in general easily separated, but there can be some overlap of characters, and care is needed in identification. The key and descriptions given in CTW are rather misleading.

1. Calyx persistent in fruit, divided at anthesis to less than 1/2 its length, teeth broadly triangular (equilateral); corolla up to 8mm in diameter; style ± equalling calyx tube; stem often with stolons from the lower axils. M. palustris (L.) Hill (M. scorpioides L.)

1. Calyx often caducous at maturity, divided at anthesis to at least 1/2 its length (or if less than 1/2 its length, then corolla 5mm or less, style 1/3 to 1/2 length of calyx tube, and stolons absent).

2. Stems with numerous spreading hairs towards the base; rooting stolons produced from the lower axils; calyx deeply divided to 1/2 to 2/3 its length. M. secunda A. Murray

2. Stems with appressed hairs (or very few spreading), or glabrous at base.

3. Stolons numerous at base of stem; leaves short and wide, normally 2 times as long as broad (rarely 2½ times); calyx deeply divided, with ± rounded teeth; corolla grey or pale blue. Wet places in upland areas. M. stolonifera (DC) Gay ex Leresche & Levier (M. brevifolia C.E. Salmon)

3. Stolons absent; leaves, even the uppermost, at least 2½ times, and normally at least 3 times as long as wide; calyx not deeply divided, teeth subacute, 1/3 to 2/5 calyx length. M. laxa Lehm. subsp. caespitosa (C.F. Schultz) Hyl. ex Nordh. (M. caespitosa C.F. Schultz)

Myosotis x suzae Domin (M. caespitosa x M. scorpioides) is a vigorous perennial, morphologically intermediate between the parents. It is very variable and partially fertile, sometimes forming complex populations. It can be identified by intermediacy in:

i) corolla size (5-7mm in diameter)
ii) style length (about equalling the developed nutlets)
iii) the long racemes with small calices, which become brown and shrivelled, and contain only 0-2 developed nutlets)
iv) pollen grains very irregular in size and shape, and partially sterile.

The weak, elongated stems often fall over, and produce vegetative shoots from the leaf axils.

The hybrid is probably widespread in marshes where the two parents grow close together, but it has not yet been seen in the region.

## MYOSOTIS DISCOLOR Pers./M. RAMOSISSIMA Rochel

A pair of species often mis-identified because of overlapping characters, or inadequate descriptions. Difficulties arise mainly later in the season when flowers are not present.

It seems unsafe to use such characters as:

i) fruiting pedicel length,
ii) calyx being 'open' or 'closed' in fruit, teeth erect or not,
iii) colour of nutlets, and whether or not truncate at base.

Nor is the nutlet character 'scarcely bordered' or 'narrowly bordered' at all easy to interpret. There may also be considerable overlap in the ratio of fruiting cyme length to the length of the leafy part of the stem. It would thus seem best to identify these species of _Myosotis_ only on flowering characters.

Corolla at first yellow or white, usually becoming blue; corolla tube eventually about twice as long as the calyx. _M. discolor_ Pers.

Corolla blue (rarely white), never yellow; corolla tube shorter than calyx. _M. ramosissima_ Rochel (_M. hispida_ Schlecht.)

## CALYSTEGIA R.Br.

Species of _Calystegia_ (excepting _C. soldanella_) may be separated as follows:

1. Bracteoles flat, rarely more than 15mm wide, not overlapping; stamens 15-30mm long; anthers 4.0-6.5mm long; corolla 30-50mm long.

2. Flowers white. Stem, petioles and peduncles glabrous. _C. sepium_ (L.) R.Br. subsp. _sepium_
2. Flowers pink. Stem, petioles and peduncles often pubescent. _C. sepium_ subsp. _roseata_ Brummitt

1. Bracteoles inflated, saccate at base, 15-38mm wide when flattened, closely surrounding and often concealing the calyx; stamens 24-40mm long; anthers 6-8mm long; corolla 40-90mm long.

3. Flowers pink with paler stripes outside. Bracteoles rarely exceeding 25mm wide when flattened. Stem, petioles and peduncles pubescent, at least when young. Garden escape. _C. pulchra_ Brummitt & Heywood (_C. dahurica_ auct.)
3. Flowers white, but frequently marked with pink outside. Bracteoles frequently exceeding 25mm in width. Glabrous. _C. silvatica_ (Kit.) Griseb.

The fertile hybrid C. x lucana (Ten.) G. Don. (C.sepium x C.silvatica) mediate between the parents, is probably not uncommon, but not yet recorded in the region. Its description is as follows:

Pedicels 30-100cm long. Corolla white (? always), 41-62mm long; stamens 20-21mm long; style and stigma 20-23mm long. Bracteoles broadly ovate 14-25mm wide when opened out, acute, obtuse or mucronate at apex, weakly cordate at base, slightly or strongly inflated, midrib very prominent especially at base, edges overlapping at each side and partially obscuring the calyx.

C. sepium subsp. roseata is rare in Britain and western in distribution. It occurs in vc.70. Intermediates between subsp. sepium and subsp. roseata and may also occur.

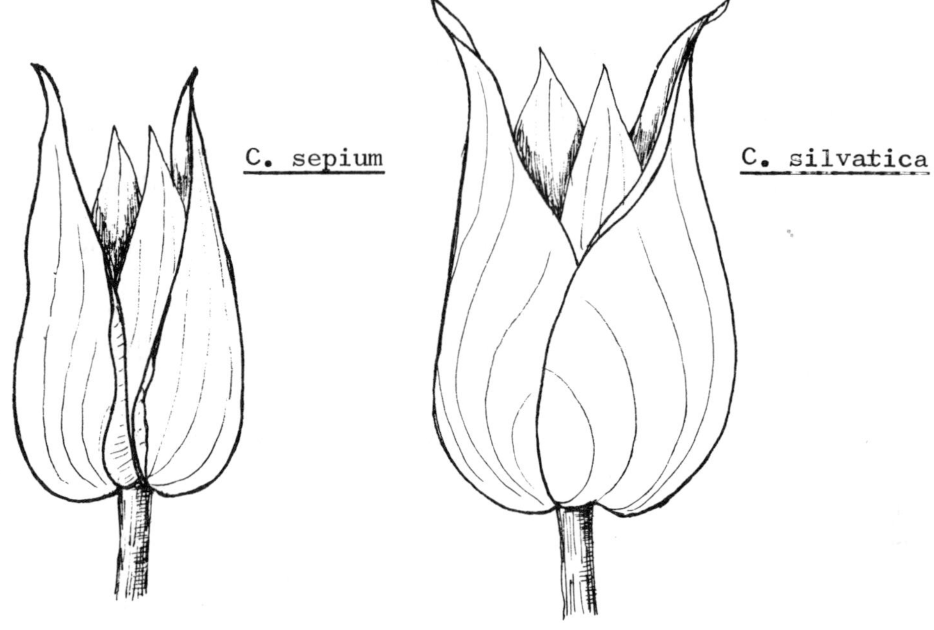


LYCIUM L.

Two introduced species, L. barbarum, and the rarer L. chinense occur in Britain, but they have not been recorded separately in our region up to the present time. They are naturalised in hedges, on walls, and in waste places.

Corolla tube narrowly cylindrical at base for about 2.5-3.0mm; corolla about 9mm in diameter; leaves usually widest about the middle.
L. barbarum L. (L. halimifolium Mill.)

Corolla tube narrowly cylindrical at base for about 1.5mm; corolla 10-15mm in diameter; leaves usually widest below the middle.
L. chinense Mill.

W.T. Stearn, in Flora Europaea, Vol.3. (1972)

LINARIA x SEPIUM Allman (L. repens (L.) Mill. x L. vulgaris Mill.)

This hybrid, apparently uncommon in our region, should be looked for wherever the parents are found growing in close proximity, for example on railway tracks, roadsides, waste ground, etc.

It is very variable, with the corolla 12-21mm, usually yellowish striped with violet. The leaves are mostly rather broader than in L. repens. It is highly fertile and backcrosses readily, producing a more or less complete range of variation between the parents.

MIMULUS L.

Mimulus Sect. Simiolus comprises aggregates of critical species originating in both North and South America. The North American complex is represented in Britain by M. guttatus, which here seems much less genetically variable than in its native country. The S. American complex comprises a group of species including M. luteus, M. cupreus and M. variegatus. This complex requires much further study and hence any treatment of British cultivated plants and naturalised taxa must remain tentative. The monograph by A.L. Grant - Ann. Missouri Bot. Gard. 2 99-388 (1924) - remains the standard account, but the treatment of M. luteus and its varieties and related species is based on inadequate material and is consequently misleading.

As far as British naturalised taxa are concerned, hybrids between M. luteus and its allies are at least partly fertile, while their hybrids with M. guttatus are effectively sterile. The hybrids of M. guttatus with M. luteus and M. cupreus have been well described by R.H. Roberts - Watsonia, 6 70-75 (1964); 6 371-376 (1968).

Sterility is used as an identification character in the following key. Where pollen is examined, the criterion of fertility is that at least 50% of pollen grains should be well formed and stainable with aceto-carmine when examined under a microscope. Sterile hybrids should have less than 20% of the pollen grains well formed; often there will be none. Staining with aceto-carmine is not essential, but more clearly distinguishes the viable pollen from the inviable. A useful but not totally reliable field test is gently to insert a finger into the throat of a mature flower. Sterile hybrids generally have indehiscent anthers and so a dusting of pollen on the withdrawn finger is a likely indication of fertility. However, at least one clone of M. guttatus x M. luteus will also release pollen. This test is not practicable on wet days.

Where mature capsules are present, presence or absence of well formed seeds provides a simple test of fertility. However, hybrids between members of the M. luteus group may fail to self-fertilise, if cross-fertilisation has not already taken place, and may set very little seed. Failure of the capsule to swell, leaving an empty non-inflated calyx, is a sign of sterility. The converse is not true and fertility cannot be judged by feel.

In using the key it should be noted that corolla patterning is variable in any taxon. This has caused the considerable confusion between M. guttatus and M. luteus. This confusion renders old literature records untrustworthy and it persists in horticultural literature to the present day. The presence or absence of glandular hairs is also of no taxonomic significance, but is determined by the prevailing weather. Sizes of all parts of the plant are readily modified by habitat conditions, hence estimates of size are best made in comparison with other locally occurring taxa (M. guttatus and M. guttatus x M. luteus may be regarded as "average"). Leaf characters are best judged from the lower cauline leaves or the summer vegetative growth.

The key given below must be regarded as provisional, and probably does not take full account of the potential variation of these very plastic plants.

1. Simple white hairs present on the inflorescence, though sometimes hidden amongst dense, long, glandular hairs; simple hairs sometimes restricted to keels and bases of the calices; plant fertile or sterile 2

   Simple white hairs absent (check calyx base with good lens) other than inside the calyx; short glandular hairs often present though sparse; plant fertile 7

2. Corolla predominantly copper-coloured, at least at maturity 3
   Corolla basically yellow, often blotched orange, red or purple 5

3. Calyx petaloid, forming apparent "double" flowers; plant relatively robust; usually sterile but rarely setting a few viable seeds. Very locally established, from Selkirkshire northwards.
   *M. cupreus* x *M. guttatus* (petaloid form)
   Calyx normal; plant sterile. 4

4. Corolla spotted but not blotched; plant slender. Locally abundant on river-shingle in N. England and Scotland. *M. cupreus* x *M. guttatus*

   Corolla with darker blotches and often tinted red; plant relatively robust. A rare, transient garden escape, or forming very distinctive local populations, or spontaneous with *M. guttatus* and *M. cupreus* x *M. luteus*. *M.* (*cupreus* x *luteus*) x *M.guttatus*

5. Throat of corolla more or less closed, spotted with red; corolla lobes unblotched (in Britain), lower lip held horizontally, much exceeding upper lip; cauline leaves typically abruptly contracted into the petiole or subsessile, often no longer than wide, teeth deltoid or becoming longer and more irregular towards the leaf-base; uppermost bracts sessile, orbicular, cuspidate, entire or very finely serrate; inflorescence tall, robust, many-flowered, clothed with simple white hairs and stalked glands in variable proportions; long stolons produced in autumn; *plant fertile*. Widespread and locally common by lowland rivers and lakes, rare and much over-recorded at altitude.
   *M. guttatus*

   Throat of corolla more usually open; corolla lobes often blotched, lower lip angled downwards; cauline leaves sometimes twice as long as wide, base often cuneate, teeth often longer than wide and twisted; uppermost bracts usually broadly lanceolate, serrate; inflorescence usually short, few-flowered, simple white and glandular hairs sparse to relatively abundant; stolons short and robust; *plant sterile*. Hybrids of *M. guttatus* (sometimes morphologically indistinguishable from *M. guttatus*). 6

6. Leaves typically longer than wide, teeth often longer than wide and twisted; blotches of corolla lobes, if present, orange to red or red-brown, infrequently covering the entire lobes, very often just one central spot on the lower lip. Widespread and locally abundant by rivers and streams in northern and western Britain. Much the most common taxon on high ground where it is usually mistaken for one or other parent. *M. guttatus* x *M. luteus*

Leaves as broad as wide, teeth even, deltoid; corollas very large, lobes heavily blotched with orange, red or purple, the entire lobes often so coloured. Difficult to separate from the last in the case of herbarium material. Still cultivated as a Scottish cottage-garden plant and locally well naturalised from Dumbartonshire northwards. M. guttatus x M. variegatus

[Synthesised plants of M. (cupreus x luteus) x guttatus are often indistinguishable from M. guttatus x luteus, or may differ only in the more diffuse blotching, hence many wild populations of this parentage may key out here. Similarly, it is very possible that the plant here regarded as M. guttatus x variegatus is, more precisely, M. guttatus x (luteus x variegatus), a hybrid that would be very difficult to confirm.]

7. Corolla in some part copper-coloured, at least where mature. 8
Corolla cream to yellow or pink, often blotched red, purple or chocolate. 9

8. Corolla uniformly copper, sometimes yellow at first; plant very small, slender, decumbent, often annual; leaves small, rhomboidal, often 3-veined, evenly and neatly toothed. Often reported as an escape, but always in error (?) for its hybrids; unlikely to establish itself in Britain and becoming replaced in cultivation by more colourful, vigorous and longer-lived hybrid cultivars. M. cupreus

Corolla very variable in colour and patterning, but typically yellow and covered in confluent small blotches of coppery orange; plant medium-sized to robust, often erect, perennial; leaves variable but usually with some teeth longer than wide and somewhat twisted. Very locally established from Peebleshire northwards. M. cupreus x M. luteus

9. Corolla yellow, throat spotted but lobes unblotched. 10
Corolla variously coloured, usually blotched. 11

10. Leaves rather small, tending to be evenly and neatly toothed. (Described above) M. cupreus x M. luteus

Leaves larger, toothing untidy, individual teeth frequently parallel-sided, to twice as long as wide, twisted. N. Scotland, rare, status doubtful. M. luteus var. luteus

11. Annual or short-lived perennial producing little lateral growth; leaves rhomboidal, hardly longer than broad, teeth shallowly deltoid; corolla cream or pale yellow, lobes largely or entirely occupied by deep pinkish-purple blotches; flowers large, typically only one open on the inflorescence at any one time. Not reported as an escape; largely replaced in cultivation by hybrid cultivars. M. variegatus

Perennial producing abundant lateral non-flowering shoots; leaves often twice as long as broad, irregularly-serrate; at least some individual teeth more or less parallel-sided, twice as long as wide, twisted; corolla pale yellow, each lobe marked with a single dark-red to chocolate blotch of variable size; flowers not conspicuously large, one to several open together on each inflorescence. A rare plant of hill-streams; much misunderstood and over-recorded; Durham and Wigtownshire northwards. M. luteus var. younganus

Populations of *M. luteus* var. *younganus* in N. Scotland, with broader leaves and larger, purplish-blotched flowers, require further investigation. Herbarium material strongly suggests hybridisation with *M. variegatus*.

Modern garden plants, sold under a variety of often worthless cultivar names, are very variable in corolla colouring and patterning. They are complex hybrids of *M. luteus*, *M.cupreus* and *M. variegatus*, often, though wrongly, called "*M. tigrinus*". Such plants have occurred as transient escapes but are unlikely to become established .

A.J. Silverside, 1978.

## VERONICA x LACKSCHEWITZII Keller
(*V. anagallis-aquatica* L. x *V. catenata* Pennell)

This hybrid has been recorded from several places in the region, and can be recognised with certainty only by its sterility. The differences between the parent species are small, but the hybrid is robust and vigorous. The flowering racemes tend to be much longer and more spreading than in either parent.

Whilst the easily recognised hybrids are sterile, in some localities, more complex hybridisation is found, and pollen can vary from 3% to 99% viability. Hybrids with high pollen viability can be very difficult to distinguish from the species.

S.M. Walters, in Stace (1975).

## VERONICA SERPYLLIFOLIA L. subsp. HUMIFUSA (Dicks.) Syme

This subspecies is found in damp places in mountainous regions. Its occurrence in our region seems generally doubtful, and plants recorded as this subspecies, are probably subsp. *serpyllifolia*.

Stems procumbent for only a short distance at base, the greater part erect; racemes elongated, with 20-40 flowers, subglabrous to eglandular-pubescent; pedicels about as long as calyx; corolla 6-8mm, white or pale blue, with darker blue veins. subsp. *serpyllifolia*

Stems often procumbent for a large part of their length; racemes usually short, with 8-15 flowers, glandular-pubescent; pedicels considerably longer than calyx; corolla 7-10mm, bright blue. subsp. *humifusa*

## VERONICA HEDERIFOLIA L. agg.

Recent work has shown that the *Veronica hederifolia* group is a polyploid complex of which 5 taxa are currently recognised. Two species occur in Britain, and of these, *V. sublobata* appears the commoner in the region.

Leaves thin, relatively bright green, rather shallowly 5-7 lobed, middle lobe as long as wide, or slightly longer than wide, ± pointed. Pedicels $3\frac{1}{2}$-7 times as long as calyx, with an adaxial row of short hairs, and usually some patent hairs outside this row, especially in distal half. Calyx glabrous or very sparsely pubescent, shortly ciliate. Corolla small, pale lilac. Anthers 0.6mm whitish. Style usually not more than 0.5mm long. Seeds suborbicular, shallowly ribbed; the somewhat red-brown, brim of the orifice distinctly smooth and shining, whitish. Stomata 26-35µm (mean 32µm). Pollen size about 32µm. Typical forms in woodland, shady places, gardens, etc.

V. hederifolia L. subsp. lucorum (Klett & Richt.) D. Hartl. (V. sublobata auct.)

Leaves thick and darker green, 3-5 lobed, middle lobe wider than long, pointed, incisions distinctly deeper than in subsp. lucorum, especially in upper leaves. Pedicels (2-)3-4 times as long as calyx, usually glabrous except for an adaxial row of patent hairs (which are longer than in subsp. lucorum). Calyx glabrous except for rather long patent cilia. Corolla relatively large, bright blue with a white centre. Anthers 1.0mm, conspicuously blue. Style 0.7-1.0mm long. Seeds large, broadly oblong to orbicular, ribbed, pale yellowish, brim of the orifice broad, partly without ridges, somewhat shining. Stomata 35-48µm (mean 42µm). Pollen size 40µm. Typical forms in arable fields, roadsides.

V. hederifolia L. subsp. hederifolia

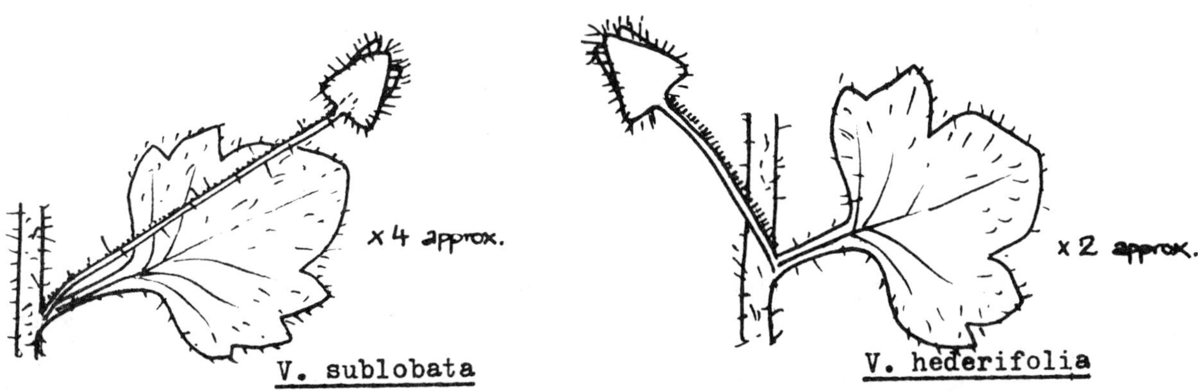


The variability of characters should be emphasised. In some plants, one or more characters may be less well shown, and plants of one subspecies may show a character which is normally associated with the other subspecies. It is essential, therefore, that all characters be assessed, and that identification be based on the combination of characters.

Anther colour appears the most constant character, whilst pedicel length is useful if taken with other features. Corolla colour is variable, as is leaf lobing. The drawings above are of typical plants in typical habitats.

M. Fisher, Ost.bot.Z., 114, 189-223 (1967), abstract in Proc.Bot.Soc.Br.Isl., 7, 435 (1968)
A.J. Silverside, BSBI News (1978)

VERONICA AGRESTIS L./V. POLITA Fr.

Both species occur widely, though uncommonly, in the region, and care is needed in their separation. The most reliable character is the type of hairs on the fruit. Examination with at least a x 10 lens is required.

Leaves all longer than broad, often lighter green and not so crowded. Corolla usually pale-blue, with lower lobe (or three lobes) white or very pale. Sepals oblong or oblong-ovate, obtuse. Fruit with long, glandular hairs, and often with short, glandular hairs, but no short crisped hairs.
V. agrestis L.

Dull green, rather crowded leaves, some broader than long. Corolla uniformly, usually deep bright blue, occasionally the lower lobe a little paler. Sepals ovate, acute or subacute. Fruit with crisped hairs, and some longer glandular hairs.
V. polita Fr.

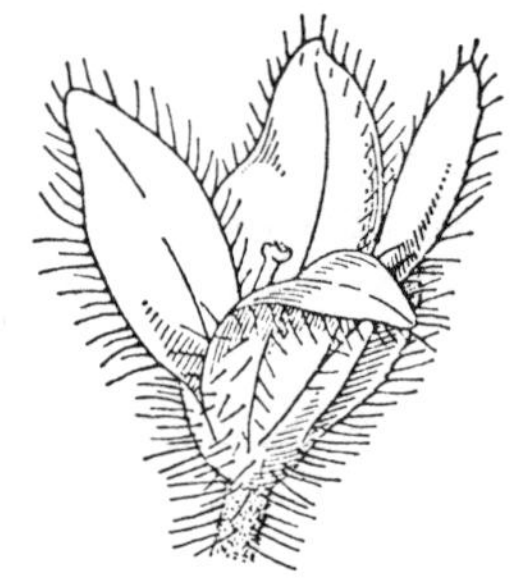

Sepals of V. agrestis V. polita

RHINANTHUS MINOR L.

Species of Rhinanthus exhibit a characteristic type of variation which appears to be mainly ecotypic in nature, and to some extent correlated with variation in the season of germination and flowering. Apart from this, the pattern of variation in other characters is so complex that there is no general agreement on the limits of the taxa.

In our area, we can separate two main lowland varieties of R. minor.

var. minor
Flowering branches (if any) suberect, mainly from the middle or upper part of the stem; leaves of main stems usually oblong; intercalary leaves (i.e. those between the uppermost branches and the bracts) 0(-1) pairs. Flowering period May to July. On the drier soils.

var. stenophyllus Schur
Flowering branches numerous, from the middle or lower part of the stem; leaves of the main stem usually narrowly lanceolate; intercalary leaves (0-)1-2(4) pairs. Flowering period July to August. On damper soils.

A third variant, sometimes named subsp. monticola (Sterneck) O. Schwarz, is often tinged purple, and has dull-yellow flowers which become treacle-brown. It occurs in upland areas, and may perhaps be found in our area.

R. de Soó and D.A. Webb in Flora Europaea, vol.3 (1972)

MELAMPYRUM PRATENSE L./M. SYLVATICUM L.

There has been considerable confusion between the two species of cow-wheat, Melampyrum pratense and M. sylvaticum. The former species is very variable and certain forms bear a close superficial resemblance to M. sylvaticum.

M. pratense

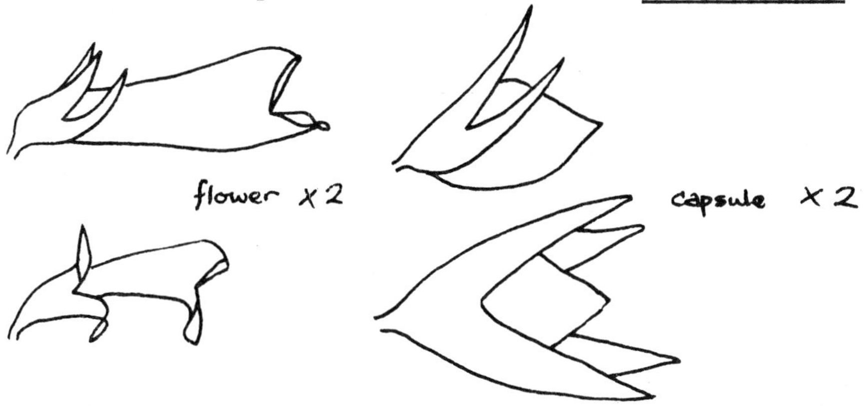

flower x 2 capsule x 2

M. sylvaticum

| | M. pratense | M. sylvaticum |
|---|---|---|
| upper bracts (2nd to 4th pair upwards ) | toothed except in some small plants having only 2-3 pairs of bracts | entire except in some large plants (usually 20cm or more in height) which may have 1-2 pairs of small teeth on uppermost bracts |
| calyx lobes | appressed to corolla | spreading |
| lower lip of corolla | not deflexed | deflexed |
| fruit | containing 4 seeds, dehiscing by a dorsal suture. | containing 2 seeds, dehiscing by a dorsal and a ventral suture. |

In the past, forms of M. pratense which superficially resemble M. sylvaticum have been recorded in error for the latter. Such forms received various names such as M. pratense var. montanum Johnston and M. pratense var. hians Druce. The names and status of these varieties may no longer be considered valid, but the taxa remain and are still being wrongly recorded as M. sylvaticum.

EUPHRASIA OFFICINALIS L. agg.

Most of the species in this complex are very variable, weakly differentiated, and hybridise readily. Populations frequently occur in which one character falls outside the normal range of variation, and hybrids are found growing with or without the parent species.

For identification, population samples of not less than six plants must be used. These should be representative of the population, ignoring extremes of variation. No allowance has been made in the key for hybrids.

In all species, especially the less well differentiated where individual characters may vary widely, it is essential that all characters be evaluated. Correct determination lies in taking the characters as a whole.

Nodes are counted from the base, excluding the cotyledonary node. Relative lengths of nodes and leaves refer to the main stem. Corolla length is measured from the base of the tube to the apex of the upper lip in its normal position (drying and pressing adds 0.5-1 mm to the length). The presence or absence of eglandular or short glandular hairs is no longer regarded as of much importance.

1a. Middle and upper leaves bearing glandular hairs with a stalk 10-12 times as long as the gland.

2a. Capsule at least twice as long as wide. Leaves often with short, and occasionally long glandular hairs. (Such plants are likely to be rare in our area, if occurring at all - see also description in 14a).
E. arctica Lange ex Rost. subsp. borealis (Towns.) Yeo
(E. borealis Wettst. and E. brevipila Burnat & Gremli)

2b. Capsule not more than twice as long as wide.

3a. Corolla not more than 7mm.

4a. Stems to 10(15)cm with 0-2 pairs of short branches; lowest flower at node (2)3-5(6); lower floral bracts 3-6(7)mm. (an exceptional small-flowered variant) E. rivularis Pugsl.

4b. Stems to 20cm. usually flexuous, with (0)1-4(6) pairs of arcuate or flexuous branches; lowest flower at node 5-8; lower floral bracts 5-12mm long. E. anglica Pugsl.

3b. Corolla more than 7mm.

5a. Lowest flower at node 2-6.

6a. Corolla 9-12mm; lower floral bracts 5-12 (20)mm.
E. rostkoviana Hayne subsp. montana (Jord.) Wettst.

6b. Corolla 6.5-9mm; lower floral bracts 3-6(7)mm long. E. rivularis

5b. Lowest flowers at node (3)6-10(14).

7a. Stem usually flexuous, with flexuous or arcuate branches; lower floral internodes usually less than $1\frac{1}{2}$ times as long as the bracts. Corolla 8mm or less. E. anglica

7b. Stem usually erect, with erect or divergent branches; lower floral internodes mostly $1\frac{1}{2}$-3 times as long as the bracts. Corolla 8-12mm.

8a. Lowest flower at node 2-6; cauline internodes 2-6(10) times as long as the leaves. E. rostkoviana subsp. montana

8b. Lowest flower at node (3)6-10(14); cauline internodes mostly not more than 3 times as long as the leaves.
E. rostkoviana subsp. rostkoviana

1b. Middle and upper leaves without glandular hairs, or glandular hairs with a stalk not more than 6 times as long as the gland.

9a. Leaves with long, whitish hairs on both sides.
E. nemorosa (Pers.) Wallr. (E. curta (Fr.) Wettst.)

9b. Leaves without long, whitish hairs.

10a. Stem to 15(20)cm , erect, stout, with erect or ascending branches; cauline internodes almost always shorter than the leaves; leaves thick and fleshy; flora bracts usually forming a dense, often 4-angled spike. Hairy forms occur on the W. coast and on the Isle of Man.Maritime habitats - cliff-tops, dunes etc.
E. tetraquetra (Bréb.) Arrond.
(E. occidentalis Wettst.)

10b. Leaves not thick and fleshy; floral bracts not forming a dense spike.

11a. Corolla more than 7.5mm.

12a. Lowest flower at node 9 or higher.

13a. Stem erect, with (0)1-9 pairs of ascending branches; cauline internodes to 4 times as long as the leaves; lowest flower at node (5)10-14; lower floral bracts mostly opposite, the basal teeth patent. Corolla 5-7.5(8.5)mm , white to lilac. Capsule usually slightly shorter than the calyx. Common on heaths, downs, pastures. <u>E. nemorosa</u>

13b. Stem flexuous, with (0)2-8(10) pairs of long, slender, flexuous branches; cauline internodes mostly less than $2\frac{1}{2}$ times as long as the leaves; lowest flower at node (2)5-12; lower floral bracts often alternate and with flowers only in alternate axils, the basal teeth apically directed; leaves near the base of the branches usually very small. Corolla 5-7 (9)mm. Capsule usually about as long as the calyx. Common in upland grassland, heaths etc. <u>E. confusa</u> Pugsl.

12b. Lowest flower at node 8 or lower.

14a. Stem erect, with 0-5 pairs of usually long, erect or ascending branches; cauline internodes 2-4(7) times as long as the leaves; lowest flower at node 4-8(10); leaves often with short, and occasionally long glandular hairs; lower floral bracts ovate to deltate, teeth usually much longer than wide. Corolla 6-9(10)mm. Capsule 4-7mm , usually not exceeding calyx. <u>E. arctica</u> subsp. <u>borealis</u>

14b. Stem and branches flexuous. See description in 13b. <u>E. confusa</u>

11b. Corolla not more than 7.5mm.

15a. Lowest flower at node 6 or higher.

16a. Cauline internodes mostly $2\frac{1}{2}$-6 times as long as the leaves.

17a. Lowest flower at node 9 or higher.

18a. Whole plant usually strongly tinged with purple. Plant slender, with (0)2-7(10) pairs of slender, erect branches. Lowest flower at node (4)6-14(16); cauline leaves narrowly-ovate to obovate, not darker beneath. Corolla 4.5-6.5mm , usually lilac to purple. Capsule usually shorter than calyx. Heaths, usually associated with <u>Calluna vulgaris</u>. <u>E. micrantha</u> Reichenb.

18b. Not with the above combination of characters. See also 13a. <u>E. nemorosa</u>

17b. Lowest flower at node 8 or lower.

19a. Plant strongly tinged with purple. See also 18a. <u>E. micrantha</u>

19b. Not with the above combination of characters.

20a. Stem erect, with 0-4 pairs of arcuate-erect branches. Cauline internodes 2-5 times as long as the leaves; lowest flower at node (2)3-6(8). Leaves light-green, often purple beneath;

lower floral bracts often alternate, the teeth acute to subacute, fairly short. Corolla (3.5)4.5-6.5mm , lower lip small, white; upper lip white or lilac. Capsule 4-5.5(9)mm, as long or longer than the calyx. Wet moorland, fens and flushes.
E. scottica Wettst.

20b. Not the combination of characters above. Teeth of lower floral bracts acute to aristate, long. Corolla usually more than 7mm. See also description in 14a. E. arctica subsp. borealis

16b. Cauline internodes mostly not more than 2½ times as long as the leaves. (Either E. confusa, E. arctica, or E. nemorosa). Return to 12a, b. *

15b. Lowest flower at node 5(6) or lower.

21a. Cauline internodes mostly less than 2½ times as long as the leaves.

22a. Stem flexuous or erect, with 0-2(3) pairs of erect branches; lowest flower at node 2-4(5). Leaves ± setose; lower floral bracts elliptical to deltate or suborbicular, teeth obtuse to subacute, usually not longer than wide. Corolla 4-7(8)mm , white to lilac, rarely purple. Capsule (4)5-7mm , as long or longer than the calyx, elliptical or ovate-elliptical, and emarginate. In the region occurring (if at all) only on the highest mountains. E. frigida Pugsl.

22b. Not with the above combination of characters. Stem flexuous or erect, with 0-8(10) pairs of branches; lowest flower at node 4 or higher. (Either E. arctica or E. confusa) Return to 14. *

21b. Cauline internodes mostly more than 2½ times as long as the leaves.

23a. Stem flexuous or erect, with 0-2(3) pairs of erect branches; lowest flowers at node 2-4(5). See also description in 22a.
E. frigida

23b. Not with the combination of characters above. Stem erect, with 0-5 pairs of erect branches; lowest flowers at node (2)3-8(10). (Either E. arctica or E. scottica). Return to 20. *

* It was found necessary to key out several species(whose characters widely overlap) at more than one place in the key. At 16b, 22b, and 23b one is referred back to a previous section. This format was adopted merely to make the key more compact, and reference back does not necessarily imply a wrong step has been taken.

The following hybrids are known in the region, but their identification is a matter for the expert.

E. confusa x nemorosa appears to be widespread. It usually has the tall habit of E. nemorosa, but shows characters of E. confusa, such as a flexuous stem and branches, with leaves less regularly opposite.

E. confusa x scottica occurs throughout the range of E. scottica in N.England. Generally intermediate between the parents, and usually found in flushed places or stream edges.

E. micrantha x E. nemorosa

E. brevipila x E. nemorosa

E. confusa x E. rostkoviana

E. rivularis x E. rostkoviana subsp. montana

E. rivularis x E. rostkoviana subsp. rostkoviana

P.F. Yeo, in Flora Europaea Vol.3 (1972), and pers.comm. (1979)

Although a key is given here to the taxa recognised by Yeo (1972), there is some doubt amongst other workers in the genus about this taxonomic treatment.

Karlsson (1976) considers that a number of the Euphrasia species recognised in Europe by Yeo (1972) would on close examination turn out to be ecotypes, more properly recognised at the infraspecific level, if at all. In his studies of Euphrasia in Sweden, Karlsson found that certain characters such as i) length of internodes; ii) node at which the lowest flower is situated; iii) number of branches; iv) shape of the leaves; v) size and shape of capsule are often habit- or habitat-correlated, and are thus unsatisfactory characters on which to base a classification.

T. Karlsson, Bot.Notiser, 129, 49-59 (1976)

ODONTITES VERNA (Bell.) Dumort

A very variable species difficult to separate adequately into infraspecific categories. There are local populations with distinct facies, and some fairly well-marked ecotypes, and the species exhibits seasonal dimorphism. The two reasonably most distinct variants are described here, but intermediates are fairly common.

Bracts longer than the flowers; branches usually shorter and suberect. subsp. verna

Bracts shorter than, or equalling the flowers; at least some branches long, and spreading at a wide angle, or sometimes nearly patent. subsp. serotina (Wettst.) E.F. Warb.

UTRICULARIA L.

All four British species of Utricularia may occur in the region, though U. australis has not yet been confirmed. Hitherto, this species was considered separable from U. vulgaris only by floral characters, but the recent paper by G. Thor suggests differential vegetative characters. Most useful appears to be the orientation of the segments of the quadrifid hairs, but we (MJW, GGG) have not examined these hairs in British specimens.

1. Leaf-segments spinose only at apices ; quadrifid hairs type 1 (fig. 2E) ; corolla less than 8mm. U. minor L.

1. Leaf-segments spinose on margins and apices ; quadrifid hairs types 2,3, or 4 ; corolla more than 8mm.

2. Leaf-bearing shoot 10-20cm ; stems of two kinds, a) bearing green leaves with or without bladders, b) colourless, bearing bladders on much reduced leaves, often buried in the substratum ; leaf-segments flat ; quadrifid hairs type 2 (fig. 3H). U. intermedia Hayne

2. Leaf-bearing shoot 20-100 cm ; stems of one kind, bearing green leaves with numerous bladders ; leaf-segments rounded ; quadrifid hairs types 3 or 4.

3. Leaf-segments denticulate, with spines on the teeth (fig. 4G) ; quadrifid hairs type 3 (fig. 4H) ; lower lip of corolla flat. U. australis R.Br. (U. neglecta Lehm.)

3. Leaf-segments ± entire, with spines (fig. 5G) ; quadrifid hairs type 4 (fig. 5H) ; lower lip of corolla with deflexed margins. U. vulgaris L.

G. Thor, Svensk Bot. Tidskr. 73 381-395 (1979)

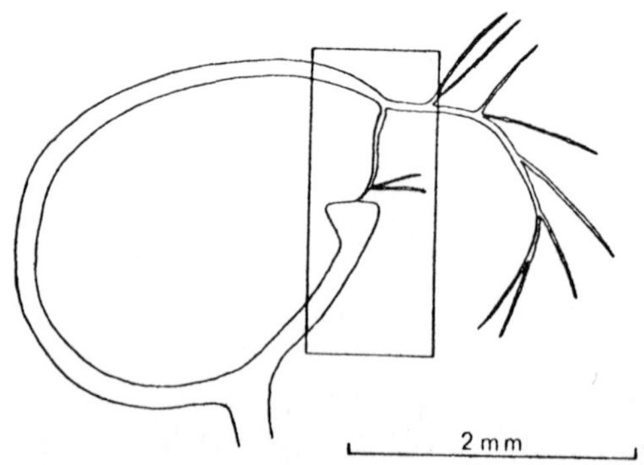


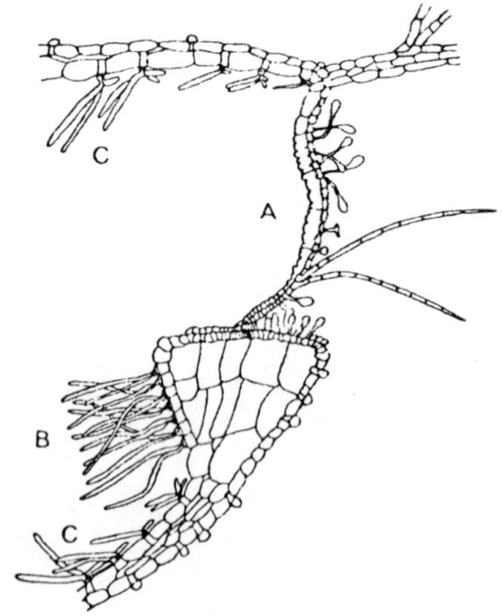


Longitudinal section through a trap of *Utricularia vulgaris*
A: Trap opening. – B: Bifid hairs. – C: Quadrifid hairs.

**Figure 1** trap = bladder in key.

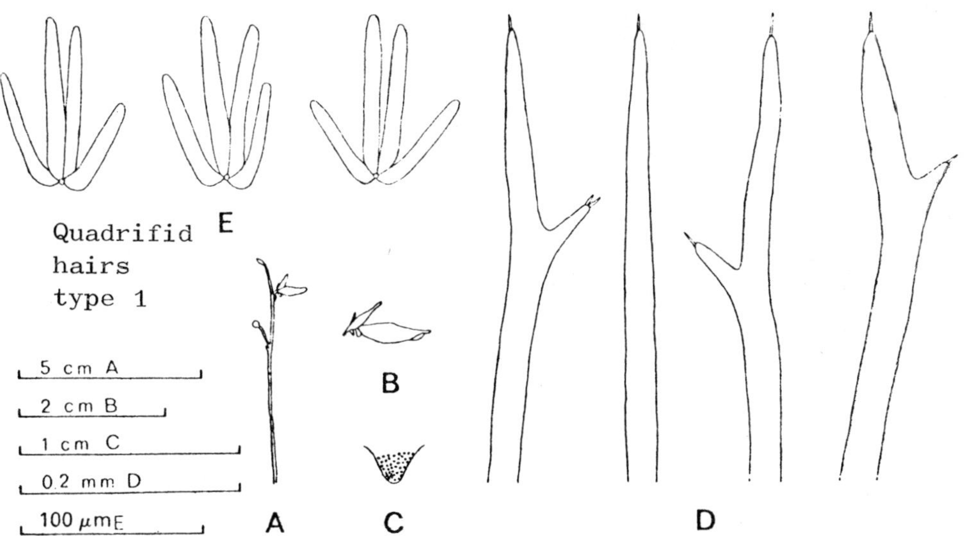


*Utricularia minor*. – A: Inflorescence. – B: Flower, side view. – C: Spur with glands. – D: Leaf segments. – E: Quadrifid hairs.

Figure 2

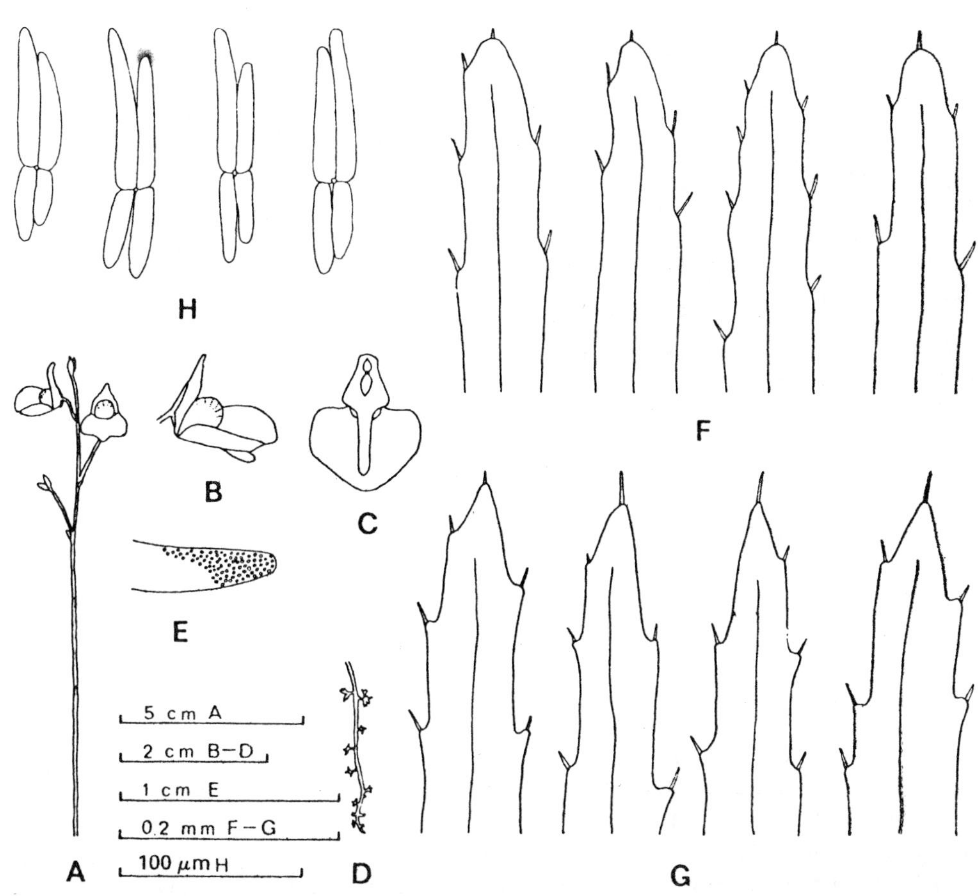


*Utricularia intermedia*. – A: Inflorescence. – B: Flower, side view. – C: Flower, back view. – D: Rhizoid. – E: Spur with glands. – F: Leaf segments. – G: Leaf segments, appearance sometimes occurring in spring and autumn. – H: Quadrifid hairs.

Figure 3 (Quadrifid hairs, type 2)

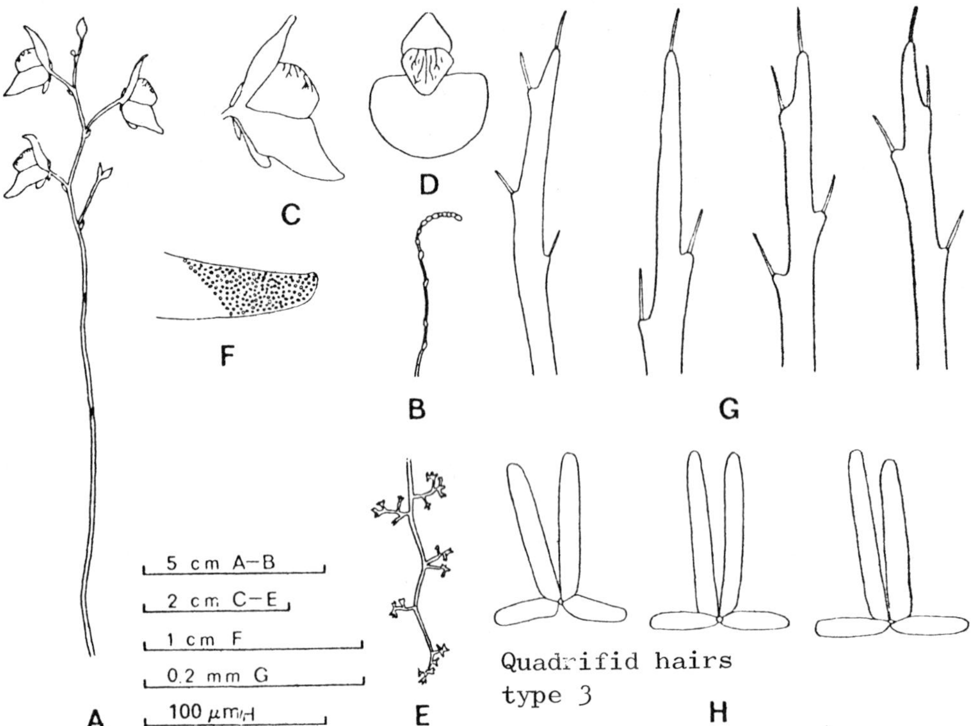


*Utricularia australis*. – A: Inflorescence. – B: Air shoot. – C: Flower, side view. – D: Flower, front view. – E: Rhizoid, characteristic appearance in Sweden. – F: Spur with glands. – G: Leaf segments. – H: Quadrifid hairs.

Figure 4

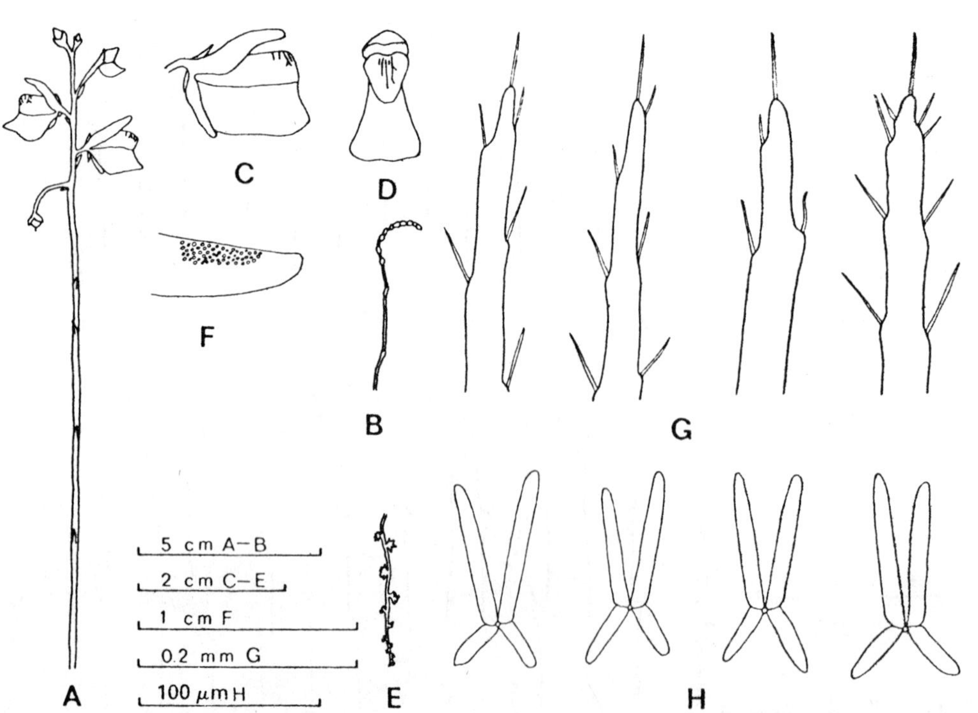


*Utricularia vulgaris*. – A: Inflorescence. – B: Air shoot. – C: Flower, side view. – D: Flower, front view. – E: Rhizoid, characteristic appearance in Sweden. – F: Spur with glands. – G: Leaf segments. – H: Quadrifid hairs.

Figure 5 (Quadrifid hairs type 4)

MENTHA L.

This genus presents particular difficulties in identification because of species plasticity, genetic variability, cultivation, vegetative reproduction, and hybridisation. For this reason, voucher specimens are required for all, except M. aquatica and M. arvensis.

Specimens should have a well-grown main stem with a well-developed inflorescence bearing flowers and fruit. Side shoots should be avoided. If hybridity is suspected (included anthers, no fruit, etc) any mints growing nearby should be noted. Note also the odour (if any) of the plant, distinguishing spearmint (carvone) from peppermint (menthone) if possible.

1. Calyx hairy in throat, with distinctly unequal teeth (the lower subulate, the upper shorter and wider); leaves small, 8-30 x 4-12mm, narrowly-elliptical, attenuate at base. Very rare in region.
Pulegium vulgare Miller
(Mentha pulegium L.)

1. Calyx glabrous in throat, with ± equal teeth; leaves larger.

2. Bracts similar to the leaves; inflorescence terminated by leaves, or by very small upper verticillasters (pseudo-whorls).

3. Plant hairy, green, usually fertile; calyx 1.5-2.5mm, broadly-campanulate, the teeth deltate or broadly triangular. M. arvensis L.

3. Plant glabrous or hairy, often tinged with red, usually sterile; calyx 2-4mm, narrowly-campanulate or tubular, the teeth narrowly-triangular to subulate.

4. Calyx 2-3.5mm, campanulate, the teeth rarely more than 1mm; plant usually glabrous.
M. x gentilis L.
(M. arvensis x M. spicata)

4. Calyx 3.5-4mm, tubular, or if shorter, then the teeth usually 1-1.5mm and the plant distinctly hairy.

5. Plant subglabrous, with sweet scent; upper bracts usually suborbicular, cuspidate.
M. x smithiana R.A. Graham
(M. aquatica x M. arvensis x M. spicata)

5. Plant distinctly hairy, with a sickly scent; upper bracts ovate to ovate-lanceolate, not cuspidate.
M. x verticillata L.
(M. aquatica x M. arvensis)

2. Bracts mostly small and inconspicuous, not similar to the leaves; inflorescence of terminal spikes or heads.

6. Leaves sessile (the lower rarely shortly petiolate); inflorescence of 1 or more spikes 5-15mm in diam. M. spicata group.

7. Stems and leaves glabrous or sparsely hairy. M. spicata L.

7. Stems and leaves hairy.

8. Leaves not much longer than broad (length: breadth ratio usually about 6:4 or 5).

9. Leaves strongly rugose, ovate-oblong to suborbicular, (15)30-45 x (10) 20-40mm, obtuse, cuspidate, serrate, though often apparently crenate since the teeth are bent under, grey- or white- tomentose beneath, hairs on lower surface branched; stamens longer than corolla. Rare in the region, and often mistaken for *M. suaveolens* x *M.spicata* hybrids.
*M. suaveolens* Ehrh.

9. Leaves variable, but usually broader than in *M. suaveolens* (up to 65mm broad), and ± sharply-toothed; stamens not longer than corolla. Extremely variable hybrid (see footnote). *M. x villosa* Huds.
(*M. spicata* x *M. suaveolens*)

8. Leaves much longer than broad, lanceolate, (30)50-90 x 15-30mm, acute or acuminate, smooth or rugose, sharply- and ± regularly serrate; hairs on lower surface both simple and branched. *M. spicata* L.

6. Leaves distinctly petiolate; inflorescence globose, or an oblong spike 12-20 mm in diameter.

10. Inflorescence an oblong spike ; leaves usually lanceolate ; plant sterile. *M. x piperita* L.
(*M. aquatica* x *M. spicata*)

10. Inflorescence globose, sometimes with one to three verticillasters below ; leaves usually ovate.

11. Leaves and calyx-tube hairy; plant fertile. *M. aquatica* L.

11. Leaves and calyx-tube glabrous or subglabrous; plant sterile.
*M. x piperita* L.
(*M. aquatica* x *M. spicata*)

*M. x villosa* Huds. (*M. spicata* x *M. suaveolens*) is an extremely variable hybrid. It may be intermediate between the parents, but is often very similar to one or the other, particularly *M. spicata*, but it usually has broader, more rugose leaves. The most widespread form is nm. *alopecuroides* (Hull) Briq., often misidentified as *M. suaveolens*. It is recognised by its more robust habit, its sweet odour (like *M. spicata*), the patent teeth on its broadly-ovate or orbicular leaves, and its robust spike of pink flowers. Widespread in the region in waste places, on roadsides etc.

It is now clear that *M. longifolia* has in the past been erroneously recorded from Britain, being mis-identified for hairy variants of *M. spicata*. However, the hybrid *M. longifolia* x *M. spicata* (*M. x villosonervata* auct.) is widely cultivated, and has become locally naturalised. It is often confused with *M. spicata* and *M. x villosa* from which it is not readily separated. It differs mainly from the latter in its narrower, usually patently-toothed leaves, with few or no branched hairs. It is sterile.

calyx broadly-campanulate (*M. arvensis*) x8

calyx tubular (*M. x piperita*) x8

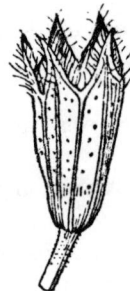

(in practice it is not always easy to decide whether the calyx is 'campanulate' or 'tubular')

R.M. Harley, in *Flora Europaea* Vol.3 (1972)

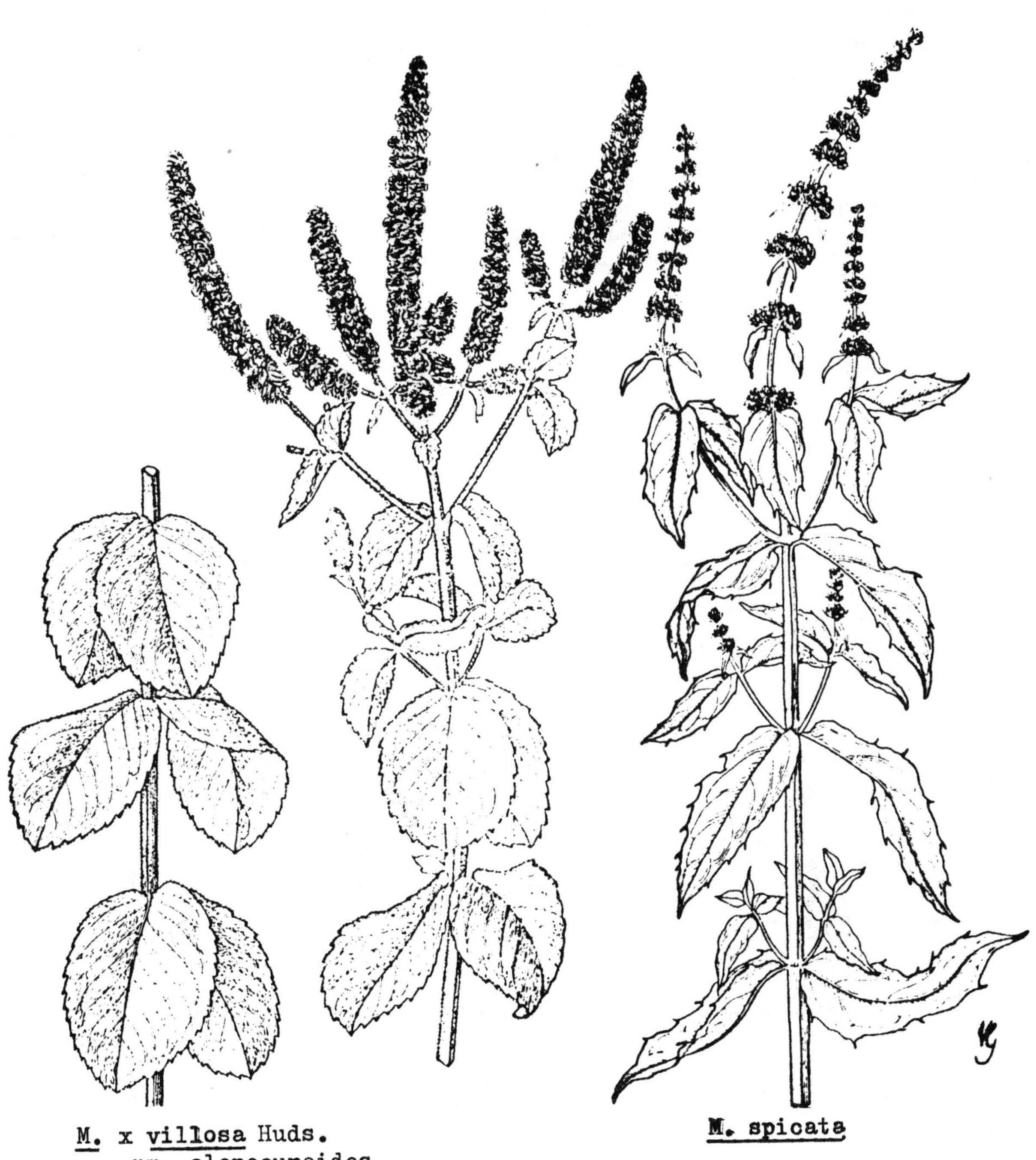

M. x villosa Huds. nm. alopecuroides

M. spicata

STACHYS x AMBIGUA Sm. (S. palustris L. x S. sylvatica L.)

The hybrid can be found in the absence of one or both parents. It is highly sterile, but care should be taken not to mistake it for male-sterile plants of the parent species. Leaves of the hybrid are intermediate in shape.

| | S. palustris | S. x ambigua | S. sylvatica |
|---|---|---|---|
| petiole length as % leaf length | 2-9% | 9-16% | 30-44% |
| nutlet production | mature nutlets produced | mature nutlets rarely produced | mature nutlets produced |
| corolla colour | pale pink | bright red | dark mauve |
| leaf length/breadth | 3.0-5.0 | 2.0-4.0 | 1.2-2.0 |
| odour | not foetid when crushed | foetid when crushed | foetid when crushed |

The hybrid is widespread in the region on riverbanks, waste ground, etc.

C.C. Wilcock and B.M.G. Jones, Watsonia, 10, 139-147 (1974)

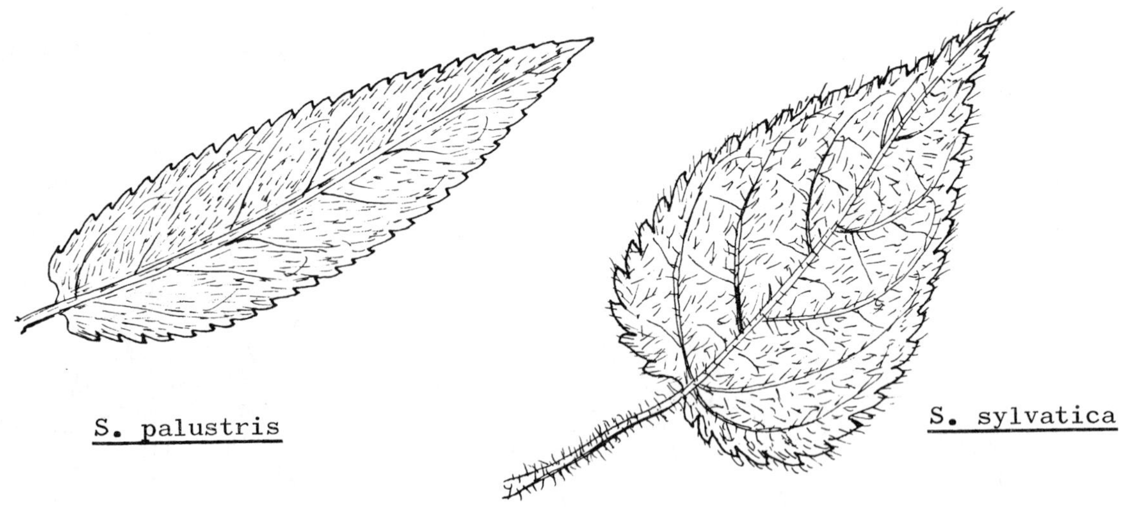

S. palustris S. sylvatica

LAMIUM L.

Some forms of L. hybridum are very similar to L. moluccellifolium The latter species normally has a larger calyx, larger lower lip of corolla, and more regular crenate/dentate leaves and bracts.

In the key given in CTW, comparisons of leaf and bract-shape, and the hairiness of the corolla-tube are difficult to interpret, and appear not to be reliable. All species, except L. amplexicaule, may have some upper bracts petiolate.

1. Calyx 8-12mm, the teeth usually longer than the tube; lower lip of corolla about 4mm. L. moluccellifolium Fries

1. Calyx 5-7mm, the teeth not longer than the tube; lower lip of corolla 1.5-2.5mm.

2. Bracts ± amplexicaul, usually wider than long; corolla usually long-exserted. L. amplexicaule L.

2. Bracts not amplexicaul (most at least shortly petiolate), longer than wide.

3. Leaves and bracts ± regularly crenate or crenate-serrate, not decurrent along petiole. L. purpureum L.

3. Leaves and bracts ± irregularly incised dentate, the upper ± decurrent along petiole. L. hybridum Vill.

GALEOPSIS TETRAHIT L./G. BIFIDA Boenn.

These two species are very similar, differing only in the shape of the corolla. Distinctions in stem hispidity and leaf dentition given in CTW are not very useful. Corolla markings are variable.

Middle lobe of the lower lip of the corolla broad and flat, entire. G. tetrahit L.

Middle lobe of the lower lip of the corlla narrower, convex with the margins ± deflexed, distinctly emarginate. G. bifida Boenn.

G. x ludwigii Hausskn. (G. tetrahit x G. bifida) can be distinguished by its intermediate corolla, its high proportion of sterile pollen grains, and few developed nutlets. It has not been recorded in our region, but may perhaps be found where the two species occur together.

GALIUM ALBUM Mill. subsp. ALBUM
(G. mollugo L. incl. G. erectum auct. angl.)

The two subspecies of G. mollugo were once distinguished on corolla size, and on characters of the pedicel and inflorescence. The variation in these taxa is now considered better summarised by placing them all in Galium album Mill. subsp. album.

GALIUM x POMERANICUM Retz.
(G. album Mill. subsp. album x G. verum L.)

Hybrids are intermediate between the parents in corolla colour, inflorescence shape, and leaf shape. It is less fertile than the parents, but some segregation and backcrossing probably occurs. The hybrid does not blacken on drying.

SAMBUCUS CANADENSIS L.

This species has been planted on railway property, and is closely allied to S. nigra.

Deciduous shrub with a bushy habit, up to 4m high, stoloniferous and clump-forming (bushes having a number of trunks); young branches glabrous; leaves bright-green, yellowish or golden (becoming greener in late summer), lower surface glabrous or slightly hairy; leaflets usually 7(sometimes 5, 9, 11), oval, oblong or roundish-ovate, the largest 15 x 6.5cm, the lowest pair frequently 2-3 lobed; teeth of the leaves small and sharp, leaf points attenuated, and in the youngest and smallest, somewhat reflexed; flowers in a strongly convex umbel 10-20cm across, white, July to September; fruit purple-black.

It may be separated from S. nigra at all seasons by the teeth of the leaves being smaller and sharper than in S. nigra, and the leaf points attenuated. The hummocky inflorescence, and the later and protracted flowering period is very characteristic. It is not unusual to see flowers and ripening fruit in late September. (S. nigra has a single flush of flowers in mid-summer). Berries are purplish-black when ripe, but very often remain greenish.

S. canadensis can be confused with S. nigra var. aurea which is also yellowish or golden, but they can be separated by the characters of foliage and inflorescence given above.

D. McClintock, and K. Hollick.

VALERIANELLA LOCUSTA (L.) Betcke subsp. DUNENSIS (D.E. Allen) P.D. Sell

Differs from the type in the much dwarfer (1-3cm), more compact, acaulescent and cushion-like habit, with leaves up to only 3.5cm long, compared with up to 7cm long in the type.

It occurs locally on west coast sand dunes, and is known in v.c.69.

GALINSOGA Ruiz. & Pav.

Two introduced species occur occasionally in the region. They differ in a number of characters, mainly floral.

| | G. parviflora Cav. | G. quadriradiata Ruiz. & Pav. (G. ciliata (Raf.) Blake) |
|---|---|---|
| plant | nearly glabrous to moderately pilose | moderately to densely pilose |
| outer phyllaries | persistent | caducous |
| receptacle scales | deeply-trifid | usually entire, or weakly bifid or trifid |
| pappus-scales of central achenes | usually ± equalling corolla and achene | usually much shorter than corolla and achene |

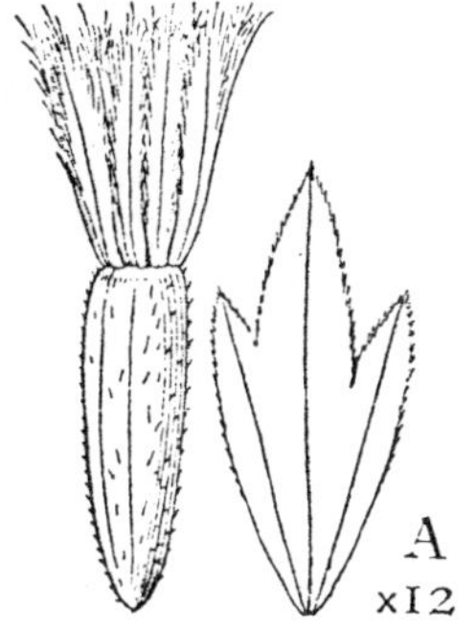


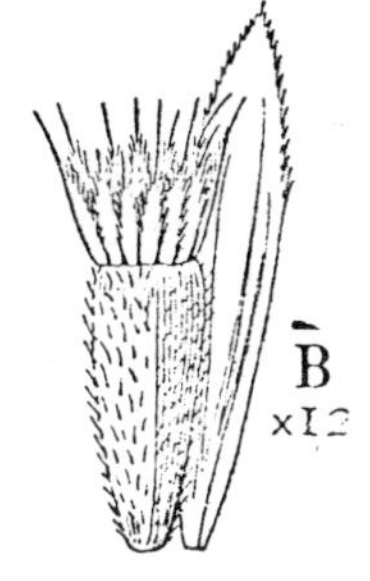


achene of receptacle scale of

A) G. parviflora
B) G. quadriradiata

SENECIO VULGARIS L. x S. SQUALIDIUS L.

Plants differing from S. vulgaris only in having ray florets are named S. vulgaris subsp. vulgaris forma radiatus Hegi. They are produced by introgression from S. squalidus into S. vulgaris. The forma occurs locally in the region, and is particularly frequent around Durham City.

The F1 hybrid (S. x baxteri Druce) is triploid, has intermediate floral characters, and mainly sterile pollen. Its presence in Britain has been proved cytologically on only three occasions. It is not known in our region. S. x baxteri is intermediate in many characters, the most obvious being in the ligule length.

SENECIO x LONDINENSIS Lousley
(S. squalidus L. x S. viscosus L.)

This hybrid is morphologically very variable, sometimes forming complex populations in which the whole spectrum from one parent to the other is seen. Hybrid plants can be recognised by their ± intermediate characters and lower fertility.

It occurs on railway ballast, waste-ground, etc., with both parent species.

P.M. Benoit, P.C. Crisp, and B.M.G. Jones, in Stace (1975)

PETASITES HYBRIDUS (L.) Gaertn, Mey. & Scherb.

The female plant of this dioecious species appears to be rare in the region.

The 'male' and female plants differ as follows:

'Male' - Heads short-stalked, 7-12mm, with 0-3 female and 20-40 sterile hermaphrodite florets. Inflorescence not markedly expanding in fruit.

Female - Heads more widely bell-shaped than the 'male', 3-6mm, with about 100 female and 1-3 sterile hermaphrodite florets. Inflorescence expanding in fruit (i.e. flower-head stalks lengthening) and stem lengthening - often exceeding 1m in height from May onwards.

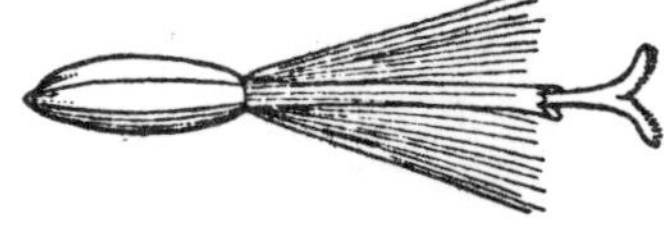

female

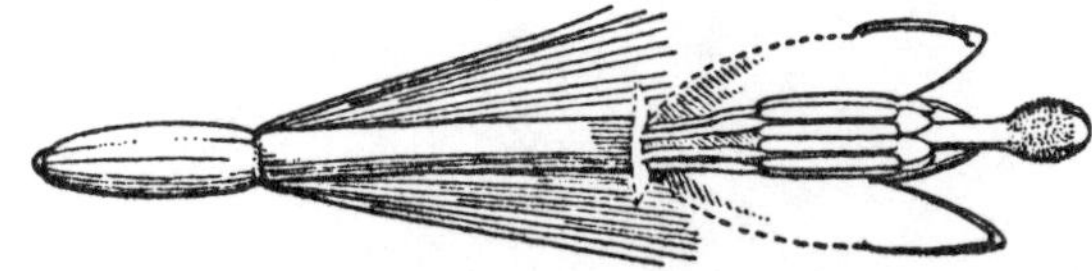

hermaphrodite ('male')

ASTER L.

Only one species and one hybrid are known in the region at the present time. They are very similar, differing only in the involucral bracts. A. novi-belgii is frequent and widespread, occurring in fens, on river banks, railway banks, and in waste places. A. x salignus is found more locally in similar places.

Outer involucral bracts tapered for most of their length, often scarious at the sides towards the base. A. x salignus Willd.
(A. novi-belgii x A. lanceolatus)

Outer involucral bracts tapered only near the apex, mainly green.
A. novi-belgii L.

P.F. Yeo, in Flora Europaea, Vol.4 (1976)

ANTHEMIS L./ CHAMAEMELUM Mill./CHAMOMILLA S.F. Gray/MATRICARIA L.

The Mayweeds present some recorders with problems, not because they are particularly variable, but because identification depends mainly upon a close examination of small floral parts.

Note that the receptacular scales subtend disc-florets, and may be present only in the central part of the receptacle. They can be 'dissected out' with a thumb-nail.

1. Receptacular scales present.

2. Corolla tube of disc-florets not saccate at base.

3. Receptacular scales narrowly-lanceolate to linear-subulate, present only in the upper half of the receptacle; ray florets usually without styles; achenes tubercled; plant glabrous or slightly hairy, foetid.
Anthemis cotula L.

3. Inner receptacular scales oblanceolate, acuminate, the outer linear-subulate; ray florets with styles; achenes strongly ribbed, not tubercled; plant sparsely to densely hairy, more aromatic. Uncommon in the region. Anthemis arvensis L.

2. Corolla tube of disc florets saccate at base, so that it covers the top of the achene; receptacular scales oblong, blunt, often laciniate at apex; achenes faintly striate on inner face; plant hairy, aromatic. No recent records in the region. Chamaemelum nobile (L.) All.

1. Receptacular scales absent.

4. Achene obovoid, with 3-5 slender ribs on the outer face, the inner smooth and lacking oil-glands; heads with ray florets soon reflexed; receptacle conical from the first, often hollow. Plant pleasantly aromatic. Chamomilla recutita (L.) Rausch.
(Matricaria recutita L.)
(Matricaria chamomilla auct.)

4. Achene turbinate, truncate above, with 3 strong ribs on outer face, the inner with 2 conspicuous oil-glands; heads with ray florets spreading until near end of flowering.

5. Annual. Achenes with well-separated ribs (by at least 1/3 of their breadth), oil-glands at upper end of inner face 1.0-1.5 times as long as broad (i.e. ± orbicular). Matricaria perforata Mérat (Tripleurospermum maritimum (L.) Koch subsp. inodorum (L.) Hyl. ex Vaarama)

5. Perennial. Achenes with contiguous or slightly separated ribs, oil-glands on inner face longitudinally elongated. Matricaria maritima L. (Tripleurospermum maritimum subsp. maritimum)

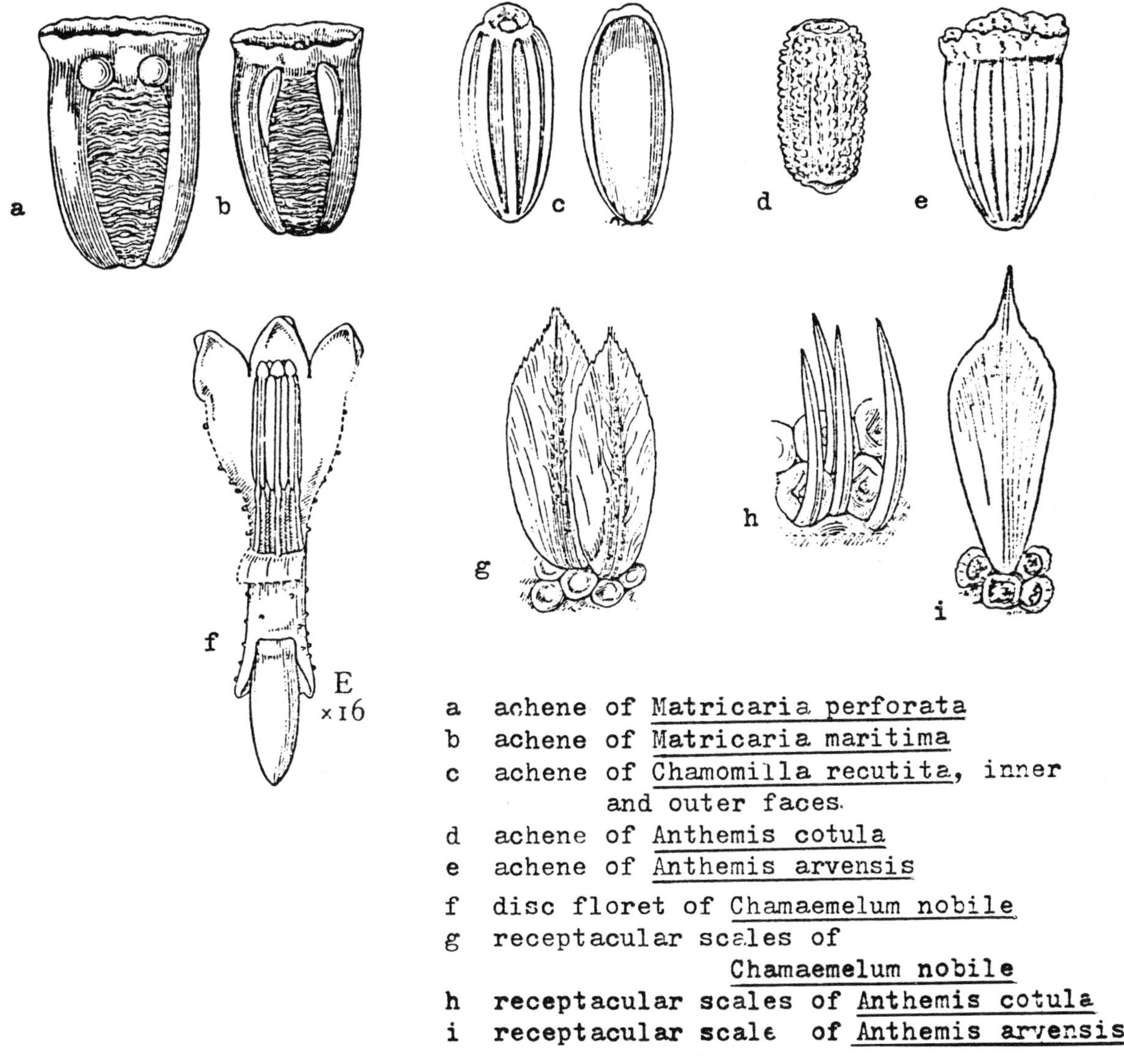


a achene of Matricaria perforata
b achene of Matricaria maritima
c achene of Chamomilla recutita, inner and outer faces.
d achene of Anthemis cotula
e achene of Anthemis arvensis
f disc floret of Chamaemelum nobile
g receptacular scales of Chamaemelum nobile
h receptacular scales of Anthemis cotula
i receptacular scale of Anthemis arvensis

ARTEMISIA VERLOTIORUM Lamotte

This introduced species has appeared in Durham in recent years, and should be looked for elsewhere in waste places, railway banks, etc.

A. verlotiorum may be separated from A. vulgaris L. as follows:

Plant tufted or with short underground shoots; stems usually glabrescent, with large central pith; leaves with only the larger veins translucent (when held to the light); inflorescence branches strict, straightish; flowering season July-September - only rarely are flowers seen later. A. vulgaris L.

Plant with long rhizomes; stems persistently pubescent, with narrow central pith; leaves with even the smaller veins translucent; inflorescence branches arcuate-divaricate; flowering season October - November - i.e., in full flower after the common species has died down, though many individual plants spread vegetatively without flowering. A. verlotiorum Lamotte

Note that:

i) In loose or sandy soil the underground shoots of A. vulgaris may be a foot or more in length, but they will be seen to radiate from a central vigorous tuft.

ii) A. verlotiorum has a much more leafy inflorescence with segments of upper leaves conspicuously elongate, linear-lanceolate, and the upper primary segments of the median and cauline leaves linear or linear-lanceolate. This gives the upper inflorescence branches a spiky appearance, as against the more compact, bushy appearance of the inflorescence of A. vulgaris. In the north, many plants never flower.

ARCTIUM

Arctium lappa L. has not been found in the region but we mention it here as some confusion has arisen owing to the continued use of the name in an aggregate sense. It can be distinguished from the other Arctium species by its large, solitary, straw-coloured heads (4.2cm in diam. including the phyllaries) on long pedicels, and by its solid petioles. The other species are not well differentiated and intermediates occur, but most plants can be assigned to one or other of the three species below.

Heads large (diameter, including phyllaries of mature heads about 3.8cm), sessile and with about 3 heads clustered at the ends of arched branches, acumens 7-10(-12)mm long; petals of open flowers not protruding from the phyllaries. The common species in our area. A. nemorosum Lej.

Heads small (diam 2-3cm) on short stalks (1cm) of primary and secondary branches; acumens 4-6(-7)mm long, petals of open flowers protruding beyond phyllaries. Rare in the region, though intermediates could occur. A. minus Bernh.

A. pubens (Bab.) Arenes is intermediate in size and with long-stalked heads (up to 15cm); acumens 8-10mm long, and involucre persistently tomentose is also worth looking for.

Note that only mature heads should be used for measurement. M.C. Lewis in the Flora of Warwickshire considers the relative lengths of the acumens of the involucral bracts to be unreliable. The acumens should be from the mid-part of the involucre. Plants growing in dense shade are difficult to identify.

## CENTAUREA NIGRA L.

There is much dispute over the limits of the two taxa C. nigra and C. nemoralis Jord. given in many floras. Dostál in Flora Europaea reduces C. nemoralis to a subsp. of C. debauxii Gren. & Godr., but the current consensus of opinion in this country is that the two taxa are not worthy of recognition as separate species and that both are best included within C. nigra. A whole range of intermediates is frequently found, and the characters used to separate them are found in varying combinations, often in a single population.

D.J. Ockendon, in Stace (1975)

## HYPOCHOERIS L./LEONTODON L.

Species of Hypochoeris and Leontodon are generally fairly distinct in their gross morphology, but the small, critical distinguishing features should always be checked (this is necessary for all atypical or stunted plants).

Hypochoeris glabra is a plant of sand dunes, and sandy fields. It is rare in the region though it may perhaps, be sometimes overlooked.

1. Receptacle with numerous lanceolate scales subtending the disc-florets. (Dissect with a thumb-nail).

 2. Pappus of 2 rows of hairs, the outer 3-6mm, the inner 9-15mm.

  3. Annual. Leaves ± glabrous; capitula 5-15mm in diameter; ray-florets about equalling the involucre, their ligules only about 2x as long as broad; inner achenes 6-8.5mm. Hypochoeris glabra L.

  3. Perennial. Leaves hispid with simple hairs; scapes usually branched and capitula usually several, 18-30(40)mm in diameter; ray-florets exceeding the involucre, about 4x as long as broad; inner achenes 8-17mm. H. radicata L.

 2. Pappus of one row of hairs, 6-11mm; leaves hispid, usually blotched with dark purple; scapes usually simple, and capitula solitary (sometimes 2-4), 18-25mm in diameter; ray-florets about 2x as long as involucre. One site in v.c. 69, and extinct in v.c.66. H. maculata L.

1. Receptacle with no bracts subtending the disc-florets.

 4. Pappus of a single row of plumose hairs; leaves glabrous, or with simple hairs; scape usually branched and bearing 2 or more heads. Leontodon autumnalis L.

 4. Pappus of two rows of hairs, the inner plumose, the outer shorter and simply-scabrid; leaves usually with forked hairs; scape simple with a single terminal head.

  5. Scape usually densely hairy above; capitula 25-40mm in diameter; outer florets orange or reddish beneath (rarely grey-violet); outermost achenes without scarious scales. L. hispidus L.

  5. Scape sparsely hairy, especially below; capitula 12-20mm in diameter; outer florets grey-violet beneath; outermost achenes surmounted by a cup of scarious scales. L. taraxacoides (Vill.) Mérat (L. leysseri G. Beck)

Leontodon autumnalis subsp. pratensis (Koch) Arcangeli differs from the common subsp. in having an involucre with dense, long hairs. It occurs mainly in upland areas, and is worth looking for in the region.

R.A. DeFilipps (Hypochoeris)
R.A. Finch & P.D. Sell (Leontodon) in Flora Europaea Vol.4 (1976)

L. hispidus x L. taraxacoides has been recorded in v.c. 67.

LACTUCA SERRIOLA L./L. VIROSA L.

There has been some confusion over the separation of L. serriola and L. virosa, arising in part from published descriptions failing to mention the two distinct leaf-types of L. serriola. They are either runcinate-pinnatifid, or unlobed, and it appears that plants with unlobed leaves are much the commoner.

The best distinction is in the ripe achene (shape, colour, and pubescence), but stem- and leaf-colours are also distinct. The two species can be identified when not flowering on a combination of vegetative characters.

| | L. serriola L. | L. virosa L. |
|---|---|---|
| overwintering rosette | leaves green | leaves grey-green, broader than those of L. serriola, often with maroon veins. |
| stem | whitish | maroon |
| stem leaves | often glaucous, with white midrib, often all arranged in the same vertical plane | green, often with maroon patches especially along veins; not all arranged in the same vertical plane |
| bracts | with spreading auricles | with auricles clasping stem |
| inflorescence | green | often tinged maroon |
| achenes | olive-grey, mottled, usually 3-4 x 0.8-1.3mm, broadest $\frac{2}{3}$ to $\frac{3}{4}$ from base (obovate), minutely ascending-hispid, and longer, paler hairs frequent towards apex | maroon to blackish, usually 4.2-4.8 x 1.3-1.6mm, rather less obovate, minutely ascending-hispid, and longer, paler hairs absent or very few |

L. serriola is a plant of S.E. England, and there is no convincing record from our region. There are few recent records of L. virosa.

S.D. Prince & R.N. Carter, Watsonia, 11, 331-338 (1977), and C. Jeffrey, pers. comm. (1975)

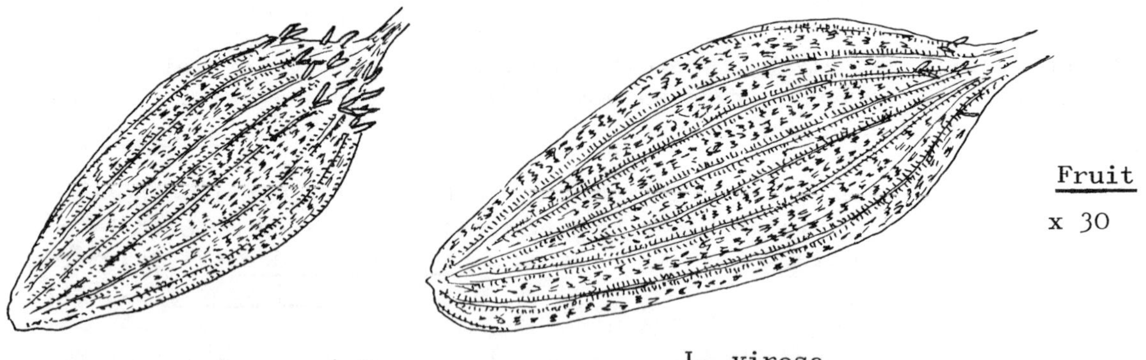

Fruit x 30

L. serriola L. virosa

HIERACIUM L.

The genus Hieracium is extremely difficult taxonomically because of its apomictic mode of reproduction (which is seen also in Rubus and Taraxacum). Apomixis results in the perpetuation of individual differences arising by mutation, by segregation, or by occasional crossing. A whole galaxy of slightly differing forms have been produced, whose features can never, or very rarely be shared.

Most of the Hieracium species in the region are local or rare, especially the upland or limestone species (Teesdale, Lake District, Arnside, etc). These species are usually well-documented, and should not be collected. Voucher specimens would be welcome, however, from lowland areas.

Notes on collecting

i) Hawkweeds should be collected at their first flowering. Species with 0-8 stem leaves should be collected not later than mid-July, and others at their first flowering.

ii) A complete plant should be collected, including basal leaves, if any, but not the rootstock.

iii) Secondary growth (i.e. after damage to the primary shoot) cannot be safely named.

iv) Plants infected by moulds, damaged by insects, or mechanically-damaged should not be taken.

v) Note in the field the colour of the leaves and styles.

The commonest species in the region appear to be H. vulgatum, H. vagum, and H. perpropinquum. The following key was originally compiled for use as a guide to lowland species in Durham, and is likely to be of only limited use elsewhere.

1. Stem leaves less than 8.

2. Stem leaves more than 2.

3. Phyllaries with numerous glandular hairs, sometimes with occasional simple hairs.

4. Leaves elliptic to ovate, ± dentate, rigid. Stem leaves 2-5 with comparatively short petioles. Phyllaries with few or no stellate hairs, with many medium to long dark glandular hairs, usually with no simple hairs. H. diaphanum Fr.

4. Leaves long-elliptic to elliptic-oblong, shallowly-dentate. Stem leaves (1-)2-3(-4) with comparatively short petioles. Phyllaries with few or no stellate hairs, with many long slender dark glandular hairs, and few simple hairs. H. diaphanoides Lindeb.

4. Leaves elliptic or ovate, regularly-dentate. Stem leaves 4-12, with rather short petioles. Phyllaries with numerous stellate hairs, with many medium and short dark glandular hairs, and usually a few simple hairs. S. England mainly, but several localities are known in our area. H. strumosum (W.R. Linton) A. Ley

3. Phyllaries with numerous simple hairs as well as a few fine short glandular hairs. Stem leaves 2-4. Common. H. vulgatum Fr.

2. Stem leaves 0-1 (rarely 2).

5. Rosette leaves thin, yellowish-green with large, $\pm$ reflexed teeth close to the cordate to rounded base. Stem leaf usually 1. All leaves $\pm$ softly-hairy above, shortly-ciliate, with villous stalks. Involucral bracts dull-green with paler margins, densely glandular but with no hairs. Styles yellow or dark. H. exotericum agg.

5. As H.exotericum, but rosette leaves dark-green, shining above. Stem-leaves 1 (sometimes 2) nearly glabrous above. Involucral bracts $\pm$ densely floccose, but not or sparsely-hairy, and usually sparsely glandular. Styles dark. H. pellucidum Laest.

1. Stem leaves usually more than 8.

6. Phyllaries glabrous, or nearly so.

7. Leaves 25-50, linear, parallel-sided with revolute scabrid margins. Outer bracts reflexed. Styles yellow. Near the coast, rare. H. umbellatum L.

7. Leaves, 20-40, ovate, ovate-lanceolate, or elliptic. Bracts subglabrous and eglandular. Styles dark. Late flowering (mid-August). Common in the region. H. vagum Jord.

6. Phyllaries with glandular hairs or simple hairs, or both.

8. Leaves (15-)25-40, denticulate; stem robust, densely long-pilose; phyllaries with long, flexuous, whitish pilose hairs, mixed with fine glandular hairs. Late flowering (mid-August). Quite common in the region. H. perpropinquum (Zahn) Druce

8. Leaves nearly always less than 15, or if not, then leaves more deeply dentate. Flowering early (beginning July).

9. Stem rather flexuous, densely pilose below; leaves 6-12, the median lanceolate with a few remote, coarse, often spreading teeth, long-attenuate below. Inflorescence of 5-6 large heads. Styles yellow-fuscous. One record in the lowlands. H. placerophylloides Pugsl.

9. Stem tall, robust, often reddish. Leaves usually less than 15, the median elliptic to rhomboid-lanceolate, sessile or shortly-stalked, with numerous sharp, unequal, ascending teeth. Heads 10-30, smaller than previous species. Styles dark. Rare in the region. H. eboracense Pugsl.

H. salticola (Sudre) Sell & West has been recorded in the region. It has leaves paler and narrower than H. vagum, often with quite large, ascending teeth, and usually with non-glandular plumaries.

HIERACIUM PILOSELLA L.

Taxa in this group are sometimes placed in a separate genus (Pilosella) but we follow Flora Europaea in placing all taxa in Hieracium.

H. pilosella subsp. trichosoma (subsp. nigrescens) appears to be the commonest subspecies in the area, especially in upland regions.

1. Ligules brownish- or purplish-red, or orange.

 2. Stolons often long and leafy, usually above ground; involucral bracts 5-8mm. H. aurantiacum L. subsp. carpathicola Naegeli & Peter (H. aurantiacum subsp. brunneocrocea)

 2. Stolons rather short, and mostly underground; involucral bracts 8-11mm. H. aurantiacum L. subsp. aurantiacum

1. Ligules yellow, often with a reddish stripe on outer face.

 3. Involucral bracts with numerous glandular hairs, without simple, eglandular hairs.

  4. Glandular hairs of involucral bracts not more than 0.5mm, more or less equal in length. H. pilosella L. subsp. micradenium Naegeli & Peter (subsp. concinnata (F.J. Hanb.) Sell & West)

  4. Glandular hairs of involucral bracts up to 1mm, very unequal in length. H. pilosella subsp. euronotum Naegeli & Peter

 3. Involucral bracts with simple, eglandular hairs, with or without glandular hairs.

  5. Involucral bracts with obvious simple, eglandular and glandular hairs.

   6. Hairs of involucral bracts pale. H. pilosella L. subsp. pilosella

   6. Hairs of involucral bracts dark. H. pilosella subsp. trichosoma Peter (subsp. nigrescens (Fries) Sell & West)

  5. Involucral bracts with dense, simple eglandular hairs, glandular hairs absent or inconspicuous.

   7. Hairs of involucral bracts pale. H. pilosella subsp. tricholepium Naegeli & Peter

   7. Hairs of involucral bracts dark.

    8. Involucral bracts and upper part of scape with moderately dense dark hairs not more than 2mm. H. pilosella L. subsp. melanops Peter

    8. Involucral bracts and upper part of scape with dense dark hairs up to 5mm. H. pilosella subsp. trichoscapum Naegeli & Peter

P.D. Sell & C. West, in Flora Europaea, Vol.4 (1976)

CREPIS L.

Owing mainly to extreme variability, the genus Crepis has always presented difficulties in determination. The following notes and key may help to overcome some of them.

i) Crepis capillaris is notoriously polymorphic. Stems may be single or several, procumbent, decumbent, semi-erect or fully erect; 1-75cm high, and 1-10mm in diam. Stem leaves are very variable. Inflorescence paniculate/corymbiform/cymose, heads 1-200 per plant, 1-3.5cm in diam. By far the commonest Crepis species.

ii) Crepis vesicaria can only be reliably identified when fully mature achenes are present (the beak elongating towards the latter part of maturity). At the juvenile stage they appear very similar to those of C. biennis. Although C. vesicaria has been known for some time in the area (e.g. since 1930 in Durham, and as a ballast alien since 1875), it is not widespread and is still being over-recorded for C. capillaris.

iii) Abortive achenes in C. biennis sometimes elongate without swelling, and give a false impression of a beak.

iv) Number of ribs on the achene (see CTW) is of use only when fully mature.

1. Involucral bracts remaining expanded after flowering; plants perennial with fibrous-rooted rootstock.

2. Stem-leaves oblong-spathulate, sessile, semi-amplexicaul; leaf margins entire or sometimes very slightly repand with a few minute, barely perceptible teeth; involucral bracts with slight scattered pubescence and gland-tipped hairs; pappus soft, white; achenes reddish-brown. C. mollis (Jacq.) Aschers.

2. Stem-leaves ovate or lanceolate, amplexicaul, cordate-auriculate; leaf margins usually with well-developed runcinate teeth; involucral bracts sparsely or thickly beset with long, black setae and short, gland-tipped hairs; pappus light-brown, brittle; achenes straw-coloured. C. paludosa (L.) Moench.

1. Involucral bracts folding inwards for a period after flowering, the tips tightening around the pappus and developing achenes, relaxing, spreading and reflexing with maturity; plants annual, biennial, or rarely short-lived perennial with short tap-roots.

3. Involucre and branches of inflorescence sparsely or thickly beset with stiff yellow bristles. C. setosa Haller f.

3. Involucre not covered with stiff, yellow bristles.

4. Involucral bracts with short, silky pubescence on inner surfaces, dark greenish-black with conspicuously light-coloured scarious margins, canescent or floccose, less frequently almost glabrous with black glandular and eglandular hairs; lowermost bracts lanceolate to narrowly-ovate, patent.

5. Stigmas greenish-brown, involucre cylindric to narrowly campanulate; heads medium-sized, 1.5-2.5cm in diam; florets sometimes with streaks of reddish-purple beneath, sometimes without; mature achenes light-brown, 3.5mm long, eventually attenuated into a slender beak of equal length. C. vesicaria L. subsp. taraxacifolia (Thuill.) Thell.

5. Stigmas yellow, involucre narrowly campanulate; heads large, 2.5-3.5cm in diam; florets not streaked with reddish-purple beneath, with a laxness in posture giving the flowers a very slightly shaggy appearance; mature achenes orange-brown 4.8mm long, tapered but not beaked. C. biennis L.

4. Involucral bracts glabrous or microscopically pubescent on inner surfaces, not conspicuously two-coloured and scarious-margined; lowermost bracts linear, appressed or slightly spreading.

6. Heads drooping in bud (? sometimes erect); involucre covered with short, light greenish-brown pubescence and occasional longer pilose hairs, sometimes also with very short gland-tipped hairs; inner involucral bracts tightly embracing marginal achenes; central achenes long-beaked, pappus persistent. C. foetida L.

6. Heads erect in bud; involucre not pubescent, sometimes with black or greenish-black glandular hairs or bristles; achenes not beaked, not embraced by involucral bracts; pappus caducous.

7. Stem closely hispidulous below; leaves slightly scabrous, sparsely or densely hispid; lower involucral bracts relaxing and slightly spreading at maturity; receptacle foveate, favose, alveoli fimbrillate (erect membranous fringes to pits, giving a honeycomb appearance beset with stiffly erect cilia); achenes orange-brown, 2.5-3.5mm. C. nicaeensis Balb.

7. Stem glabrous below, sometimes with a few scattered hairs; leaves smooth, glabrous or sometimes with scattered hairs on the midrib, and on one or both surfaces; lower involucral bracts remaining appressed (a few sometimes wither and twist a little at maturity); receptacle usually with only a few short cilia, becoming mammose; achenes light fuscous-brown, 1.5-2.5mm long. C. capillaris (L.) Wallr.

J.B. Marshall, Proceedings BSBI, 4, 393-403 (1962), 5, 325-333 (1964)

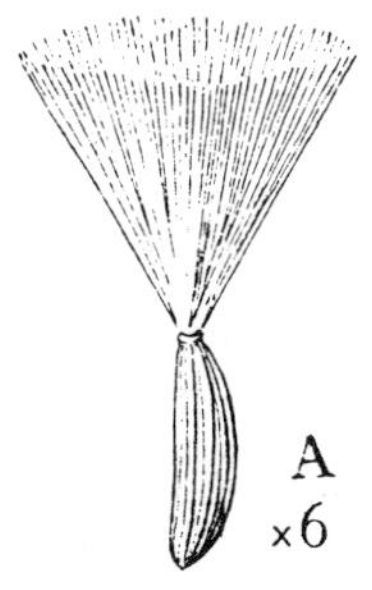


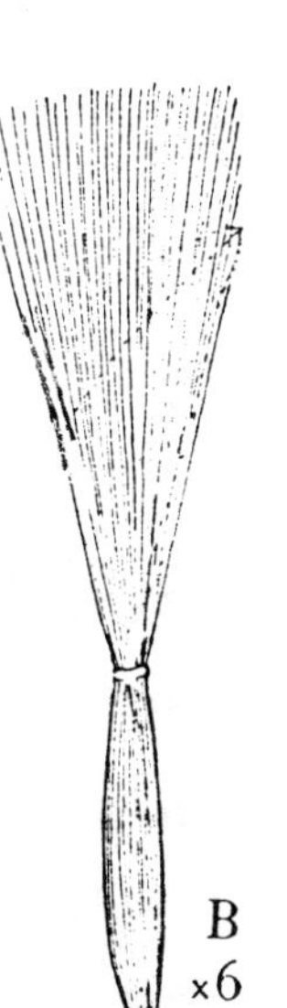


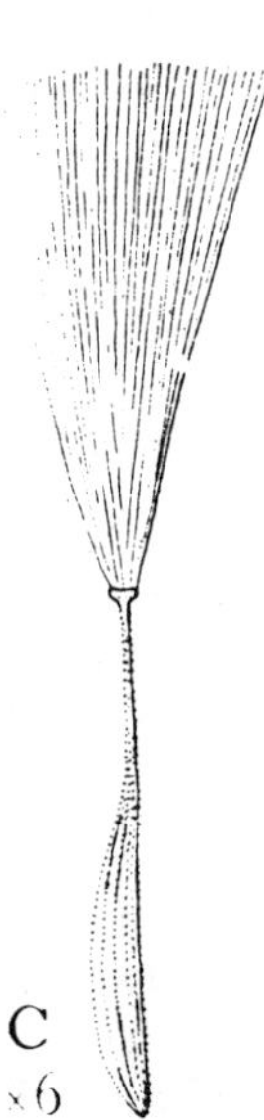


achenes of

A Crepis capillaris
B Crepis biennis
C Crepis vesicaria

## TARAXACUM Weber

The dandelions of Durham and Northumberland are quite well-documented, but there is still a need for more information on the species distribution and ecology. Cumbria has been much less well covered, and the species list is still incomplete.

Vouchers would be welcomed by the local recorders, providing the plants are well collected, and documented.

### Notes on collection

It is most important that the plants are collected carefully and selectively if identification is to be possible.

i) Collect until late-May in the lowlands, and until late-June in upland areas. Leaves produced later will be the 'summer leaves', and are usually larger and not of typical shape.

ii) Plants should be well-grown, and not from shaded, heavily-trodden, mown, or grazed areas. Ripe achenes are required for species determination. The oldest flowers in a specimen usually go to seed in the press.

iii) Note in the field
The attitude of the outer involucral bracts (i.e. whether erect, spreading, or recurved)
The length & width of the involucral bracts, their colour (all green, suffused purple, red-tipped etc), and whether bordered.
Leaf colour (shade of green), and whether blotched or spotted.
Petiole colour, midrib colour.

iv) The whole plant should be collected (single leaves and heads are not of much use), but specimens should be excised at the top of the root. If possible include both flowering and fruiting heads (see ii) above)

v) Plants are best preserved if pressed immediately after collection (i.e. in the field) as the leaves curl very rapidly. Leaves must be flattened individually, and heads pressed from the side. Small leaves and buds in the rosette may be removed.

vi) Rapid drying is essential if the true colours are to be preserved. This will mean a change of paper at least once a day, preferably twice, and the press near a source of gentle heat (e.g. a radiator).

vii) If leaves become curled and flaccid, the whole plant may be immersed in water for up to 24 hours until it is again fully turgid. It is then easily pressed (after gently shaking to remove excess water), but additional changes of drying paper are required.

It is worthwhile attempting to place the species in the appropriate section, but note the overlap between Spectabilia and Vulgaria.

The key is taken from A J Richards' Taraxacum Flora, Watsonia Suppl. (1972).

1. Plant small, delicate, with strongly dissected leaves; exterior bracts usually corniculate (plants of dry places).

2. Achenes pale-brown; leaf lobes 6-10; capitula deep-yellow or orange-yellow, often involute. Plants of dune-slacks etc. Sect. Obliqua Dahlst.

2. Achenes variously coloured (reddish, purple, cinnamon, chestnut, browns, straw-coloured); leaf lobes 3-6(10); capitula pale yellow or yellow, rarely involute. Sect. Erythrosperma Dahlst.

1. Plant more robust; exterior bracts never corniculate, though sometimes with a small callus.

3. Leaves linear, smooth, green, entire or lobed; exterior bracts adpressed, linear to ovate-lanceolate, with a broad, scarious border. Plants of fens, very rare, and not known in our region. Sect. Palustria Dt.

3. Leaves never linear, usually lobed; exterior bracts adpressed to recurved, linear to ovate-lanceolate, never with a broad, scarious border.

4. Leaves spotted or unspotted, often dark, with a red petiole and midrib, sometimes green throughout; exterior bracts to 10mm, ovate-lanceolate to lanceolate, spreading to adpressed; ligules often striped purple or reddish; pollen often absent; achenes 3.0-5.0mm. Plants usually of wet places. Sect. Spectabilia Dahlst.

4. Leaves never spotted; exterior bracts 7-18mm, erect to recurved, lanceolate to linear; ligules usually striped grey-violet, sometimes purple, never carmine. Pollen usually present; achenes to 3.5mm. Sect. Vulgaria Dahlst.

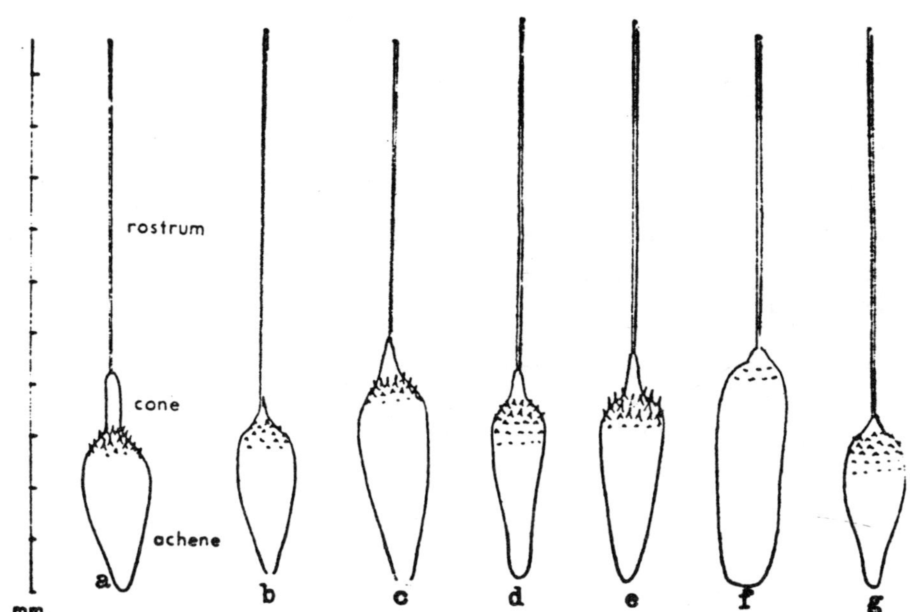


Basic fruit-shapes in Taraxacum

a) Erythrosperma
b) Obliqua
c) Palustria
d) Spectabilia (praestans group)
e) Spectabilia (crocea group)
f) Spectabilia (spectabile gp.)
g) Vulgaria

ALISMA L.

A. lanceolatum differs from A. plantago-aquatica in a number of characters, of which the floral and fruiting are the most reliable. Leaf-shape is normally distinct, but young leaves of A. plantago-aquatica can approach those of A. lanceolatum in shape.

Leaves ovate, rounded to subcordate at base; inner perianth-segments rounded; style arising below middle of carpel; anthers about 2x as long as wide. A. plantago-aquatica L.

Leaves lanceolate, gradually narrowed at base; inner perianth-segments pointed; style arising above middle of carpel; anthers about 1x as long as wide. A. lanceolatum With.

Carpels

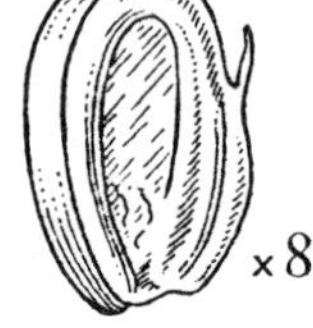

A.

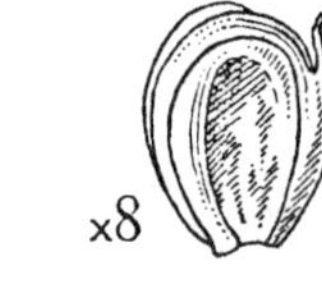

B.

A: A. plantago-aquatica
B: A. lanceolatum

ELODEA Michx.

The status of Elodea in Britain is gradually being resolved. It now seems probable that there are three species growing in this country ; E.canadensis Michx., E. nuttallii (Planch.)St John, and E. ernstae St John. E.canadensis and E. nuttallii are found in our region, and E. ernstae is confined to s. England. E. nuttallii has replaced E. canadensis at many sites and is showing no signs of slowing its rapid increase. It is more tolerant of higher nutrient levels, and this may be a factor influencing its rate of spread. All species show considerable morphological variation, and this has caused much confusion. Elongate forms of both E. canadensis and E. nuttallii are frequent in the north-west, especially in the Cumbrian Lakes. The 'Elodea nuttallii' of Esthwaite Water is now known to be Hydrilla verticillata, but it may well be extinct from this, its only known British locality.

Leaf length and width are poor characters for identification, since they show great variability. Microscopic characters, however, do prove useful, and the species can be separated vegetatively on a combination of micro- and macroscopic features.

1. Median and upper leaves in whorls of (3)4-5 ; nodal scales fimbriate. Hydrilla verticillata (L.f.)Royle

1. Median and upper leaves in whorls of 3 ; nodal scales entire.

2. Median and upper leaves usually firm, rarely reflexed, elliptic to linear-oblong, the apex obtuse ; leaf-margins with 3-10 rows of hyaline cells ; mean leaf-tooth length 60-70(72) µm ; sepals of female flowers 2.0-2.8mm long. Elodea canadensis Michx.

2. Median and upper leaves flaccid or firm and strongly reflexed, linear-lanceolate, the apex acuminate ; leaf-margins with 1-2 rows of hyaline cells ; mean leaf-tooth length 73-80 µm ; sepals of female flowers 1.0-1.8 mm long. Elodea nuttallii (Planch.)St John

D. Simpson, University of Lancaster, pers. comm. 1980

Leaf of E. canadensis

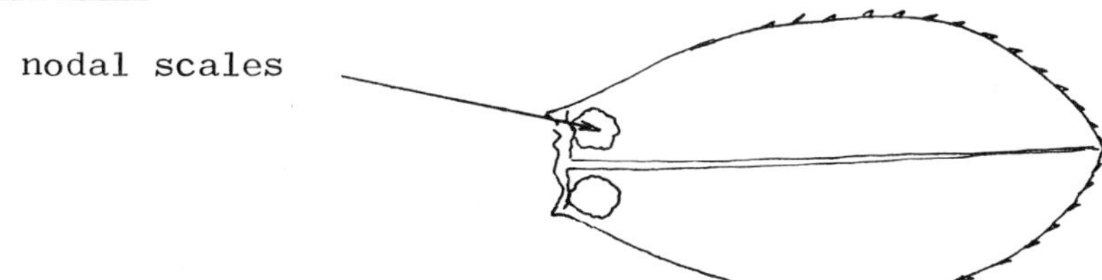


ZOSTERA L.

Zostera species grow on fine gravel, sand, or mud in the sea, from half-tide mark down to about 4m, the three species varying somewhat in their preferred range. All three species are very local in the region, and Z. marina in particular has shown a great decrease in recent years. Because of the comparative inaccessibility of the habitat, Zostera has been somewhat neglected by recorders. (Plants are, however, sometimes washed inshore by wave-action).

Many characters overlap in size, and identification should be made on a combination of characters.

| | Z. noltii Hornem. | Z. angustifolia (Hornem.) Rchb. | Z. marina L. |
|---|---|---|---|
| rhizome diameter | 0.5-1.0 mm | 1-2 mm | 2-5 mm |
| rhizome cortex | vascular bundles in innermost layers | vascular bundles in outermost layers | vascular bundles in outermost layers |
| leaves of sterile shoots | (4)6-12(20) cm x 0.5-1.0 mm | in summer 15-30cm x 2.0 mm; in winter 5-12cm x c.1mm. | 20-50(100) cm x (2)5-10 mm |
| leaves of fertile shoots | | 4-15cm x 2-3mm | shorter & narrower than on sterile, sometimes emarginate |
| leaf apex | emarginate | rounded when young, later emarginate | rounded and mucronate |
| sheaths | open | closed | closed |
| flowering stems | unbranched, or with 1-2 branches from v. near base | 10-30cm, branched | to 60cm, branched |
| inflorescence | 3-6cm | 8-11cm | (4)9-12(14) cm |
| retinaculae | present | absent | absent |
| stigma | shorter than style | approx. equalling style | twice length of style |
| seed | 2.0mm, smooth | 2.5mm, ribbed | 3.5mm, ribbed |

(retinaculae are bract-like structures, which are reduced perinths of the male flower - see G overleaf)

/drawings overleaf

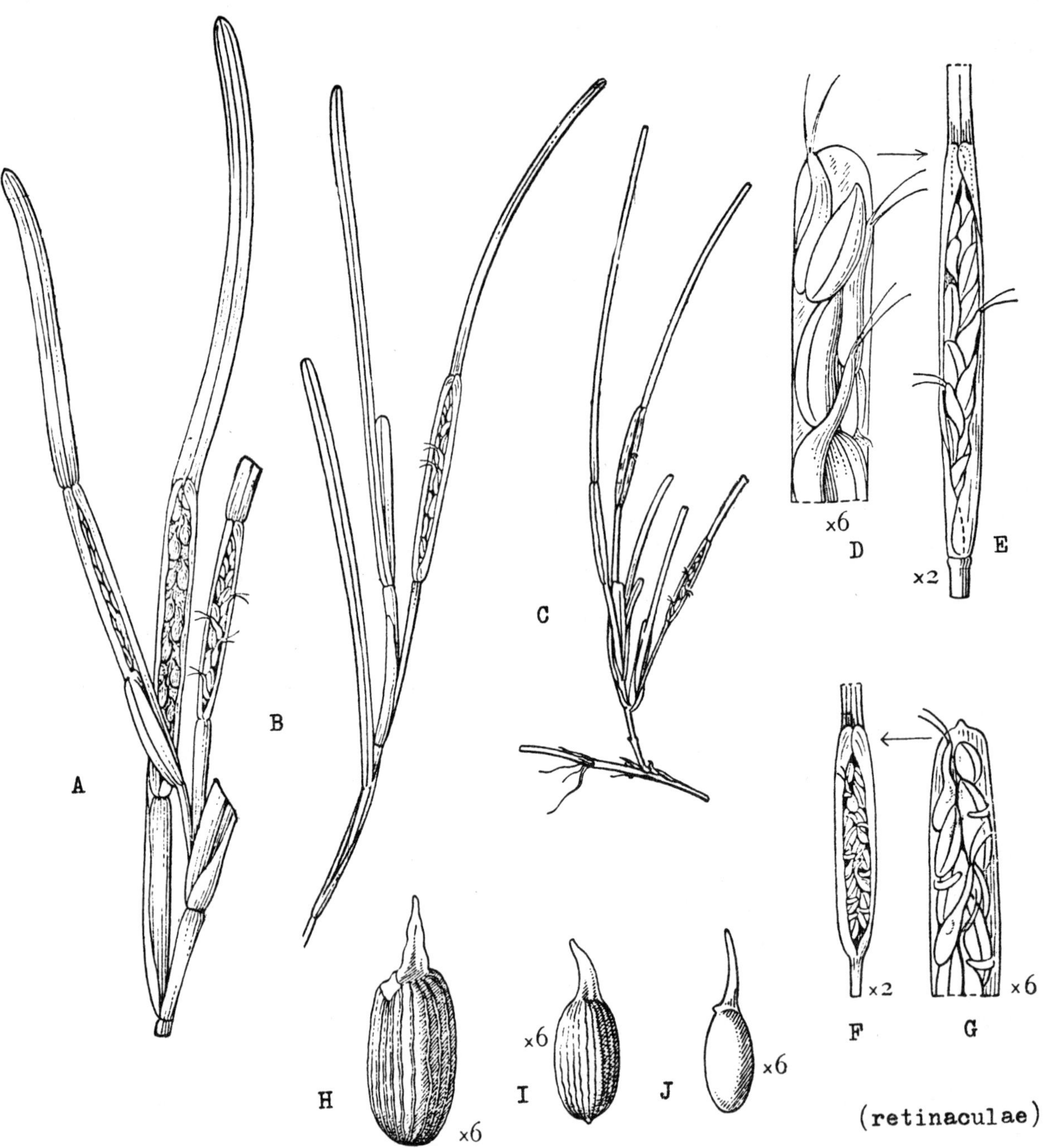


| | | | |
|---|---|---|---|
| Zostera marina | A) plant | D,E) inflorescence | H) fruit |
| Z. angustifolia | B) plant | I) fruit | |
| Z. noltii | C) plant | F,G) inflorescence | J) fruit |

## POTAMOGETON L.

Clapham in CTW notes that:-
the genus is taxonomically difficult because of the great plasticity in vegetative morphology (very different forms are produced in differing conditions of light-intensity, water-depth, rate of water-flow, nutrient-supply, and so on), and because of hybridisation. Some hybrids are locally quite common, and are usually sterile and intermediate between the parents.

Amongst the most important diagnostic features are:

i) the venation of the leaves, including the point at which the lateral veins join the midrib , and the angle of the join.

ii) the shape of the leaf apex

iii) the denticulation or otherwise of the leaf margin.

iv) whether the stipules of the young leaves are open throughout their length, or tubular in their lower part.

It is desirable that good specimens be collected from every population encountered. Of great importance is the manner of collection since crucial features may be lost or altered in poorly collected material.

A 'drag' is useful for deep-water plants. Specimens should be kept in polythene bags for as short a time as possible. Individual plants should be floated out on muslin (or other suitable material) which is then placed between constantly renewed sheets of absorbent paper to dry.

## POTAMOGETON NATANS L. / P. POLYGONIFOLIUS Pourr.

These species are very variable and can resemble each other. The petiole and stipule provide the best vegetative differentiating characters. The petiole of P. natans has a distinct twist or 'kink' just below the blade, and also has two short, narrow, decurrent wings (see Fig A below). No other Potamogeton species has this character. By contrast, the leaf of P. polygonifolius is not at all decurrent, and there is never a 'kink' in the petiole.

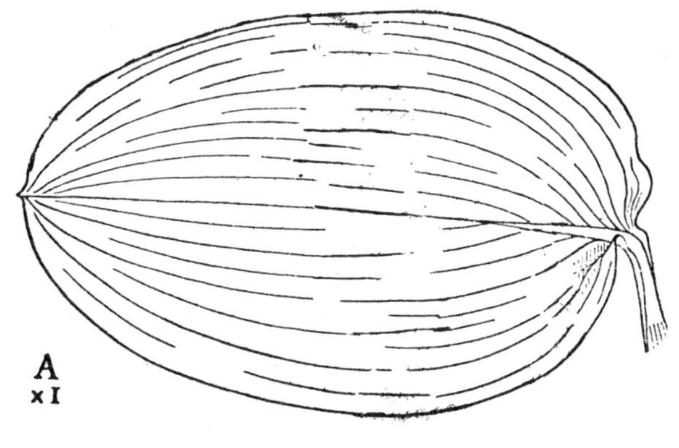


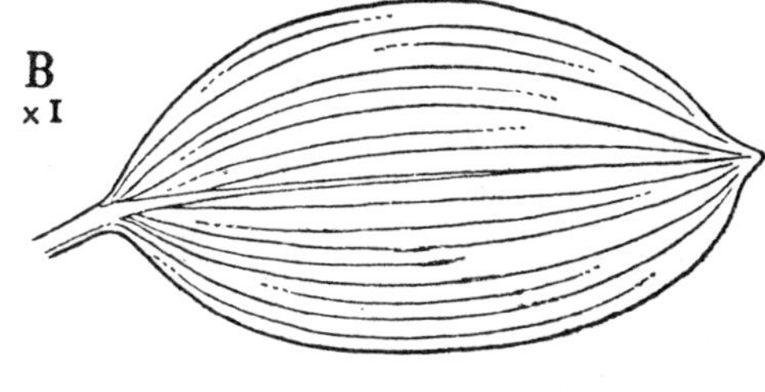


A) Potamogeton natans
B) P. polygonifolius

| | P. natans | P. polygonifolius |
|---|---|---|
| stipule | 5-12(18) cm | 2-4cm |
| fruit | 4-5 x 3 mm | 2.0 x 1.5 mm |

POTAMOGETON BERCHTOLDII Fieb./P. PUSILLUS L.

These species are variable in leaf length, width, and venation, but can be differentiated on a number of characters. The most constant are the nodal glands which are present in P. berchtoldii, but not in P. pusillus. These take the form of small swellings at the junction of the leaf-blade and the stem or branch, and can be seen with a x20 lens. (Note that this feature is given erroneously for P. pusillus in Ross-Craig).

P. berchtoldii normally has lacunae bordering the midrib of the leaf (sometimes only towards the leaf base), has stipules open from the first, and terminal turions. P. pusillus has no lacunae, stipules tubular below at first, but soon splitting, and axillary turions. The leaf veins tend to meet the midrib at a wide angle in P. berchtoldii, and at a narrow angle in P. pusillus. However, venation can be very obscure even when viewed with a binocular microscope.

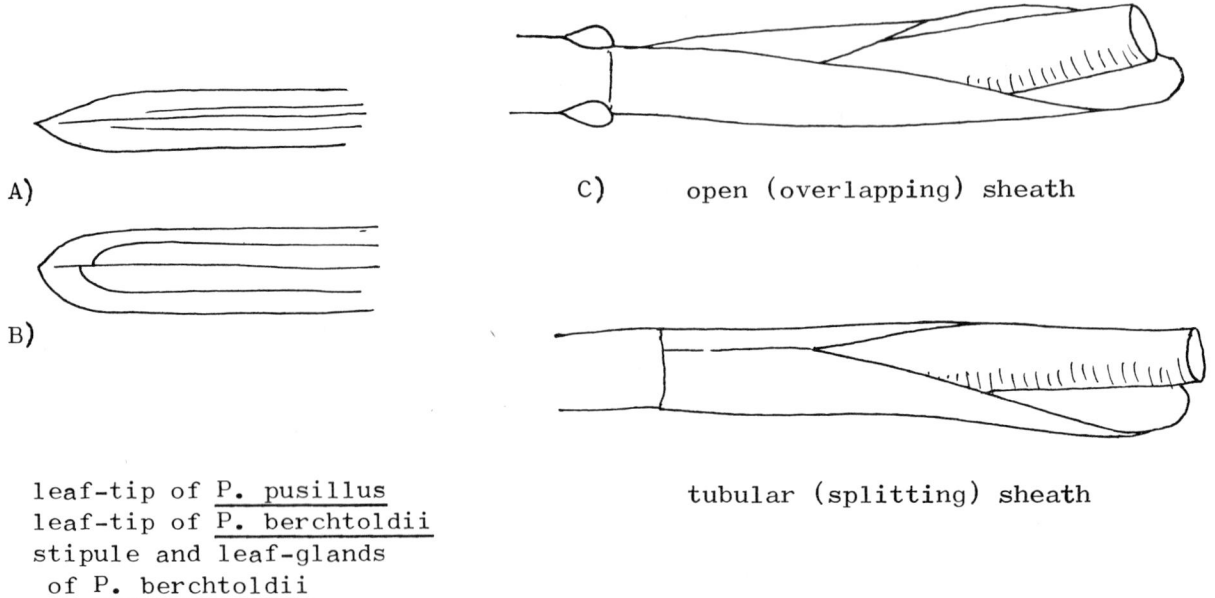

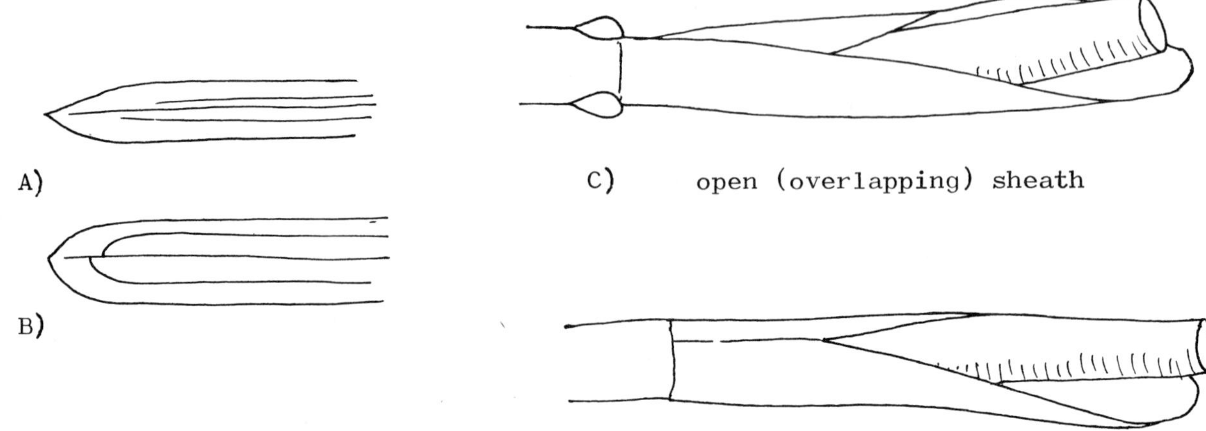


A) leaf-tip of P. pusillus
B) leaf-tip of P. berchtoldii
C) stipule and leaf-glands of P. berchtoldii

A complete list of Potamogeton species and hybrids occurring in the region is included in the Appendix. Most are local and/or rare.

NAJAS L./RUPPIA L./ZANNICHELLIA L.

Species of these three genera have narrowly-linear leaves, and can appear similar at a superficial glance, and can perhaps, also be confused with linear-leaved Potamogeton species. They are, however, readily distinguished when fruiting.

Ruppia species are found in brackish ditches and salt-marsh pools near the sea; Zannichellia occurs both in brackish and freshwater ditches and pools, and in streams and rivers. Both are scarce and local in the region. Najas flexilis is known from one lake in v.c.69, but may perhaps occur elsewhere in the Lake District.

1. Flowers in $\pm$ compact heads; fruiting carpels on long stalks.

2. Peduncle in fruit 10cm or more, many times longer than the pedicels of the carpels. Ruppia cirrhosa (Petagna)Grande (R. spiralis L. ex Dum.)

2. Peduncle in fruit 0.5-5.0cm, shorter than, to twice as long as the pedicels of the carpels. R. maritima L.

1. Flowers axillary, solitary or in few-flowered clusters; fruiting carpels not on long stalks.

3. Leaves narrowly-linear, entire; flowers 1-5 in leaf axils; carpels 2-6; fruit ± toothed on dorsal margin, with long persistent style. Zannichellia palustris L.

3. Leaves narrowly-linear, spinulose-serrulate (sometimes minutely so); flowers 1-3 in leaf axils; carpel solitary; fruit narrowly-ovoid, smooth, with shorter style. Najas flexilis (Willd.)Rostk. & Schmidt

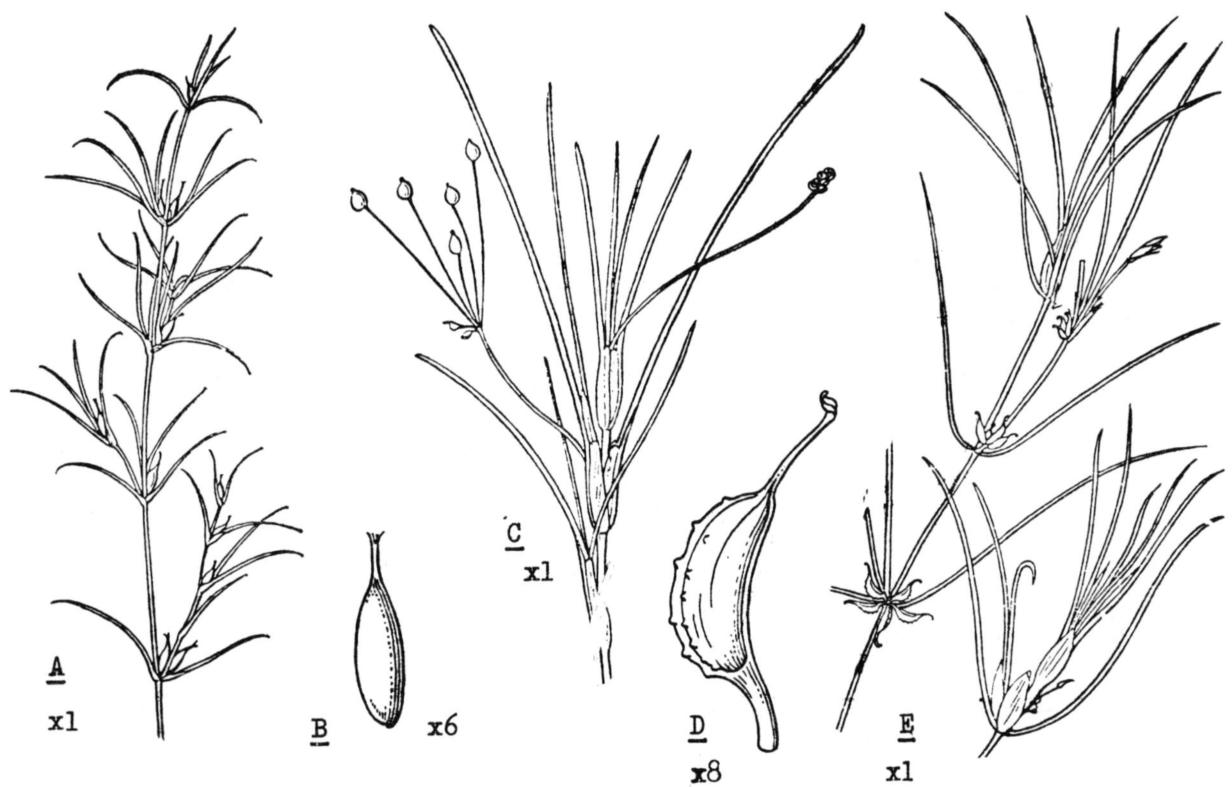


Najas flexilis A) plant B) fruit
Ruppia maritima C) plant
Zannichellia palustris E) plant D) fruit

ALLIUM L.

In general, species of Allium are not difficult to identify, but careful examination of the floral parts is required for most species.

1. Leaves ± elliptic, stalked. A. ursinum L.

1. Leaves linear or cylindric.

2. Scape terete; perianth 12 mm or less.

3. Inflorescence of bulbils only, without flowers.

4. Spathe usually 1-valved and caducous, shorter than inflorescence. <u>A. vineale</u> L. var. <u>compactum</u> (Thuill.)Boreau

4. Spathe 2-valved with long leaf-like points. <u>A. oleraceum</u> L.

3. Inflorescence with flowers.

5. Inner filaments divided at apex into 3 long points, the middle one bearing the anther, outer filaments entire.

6. Leaves flat, solid; spathe 2-valved, present at flowering; perianth reddish-purple; stamens included. <u>A. scorodoprasum</u> L.

6. Leaves cylindric or subcylindric, hollow; spathe usually 1-valved and caducous; perianth pink or greenish-white; stamens exserted. <u>A. vineale</u> L.

5. Filaments all entire, or the inner with 2 small teeth at base.

7. Spathe shorter than flowers; perianth segments 7-12mm. <u>A. schoenoprasum</u> L.

7. Spathe with long leaf-like points, much longer than flowers; perianth segments 5-7mm.

8. Anthers included; leaves semi-terete,hollow at least below. <u>A. oleraceum</u> L.

8. Anthers conspicuously exserted; leaves flat, grooved. <u>A. carinatum</u> L.

2. Scape triquetrous; perianth 10-18 mm, white.

9. Leaves 2-5; inflorescence without bulbils. <u>A. triquetrum</u> L.

9. Leaves solitary; inflorescence with bulbils. <u>A. paradoxum</u> (Bieb.) G.Don

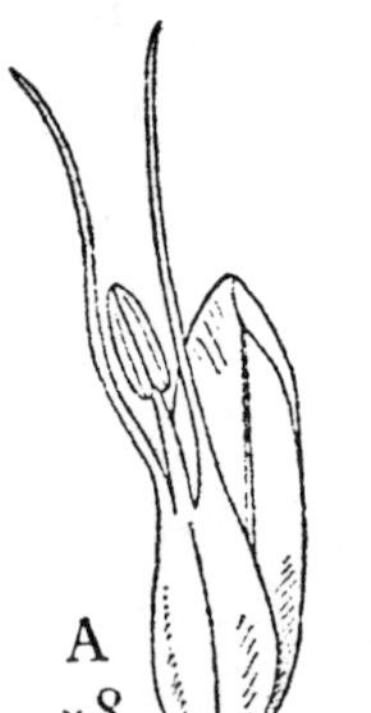


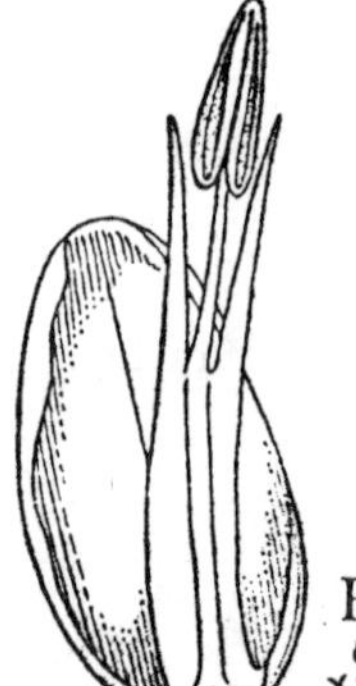


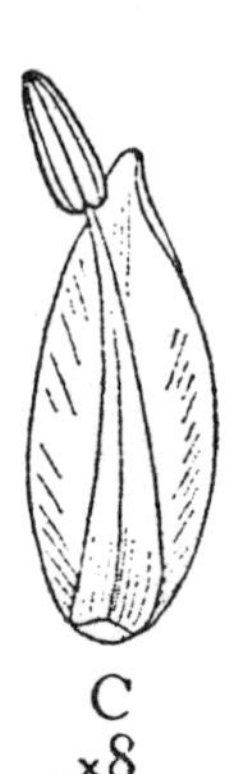


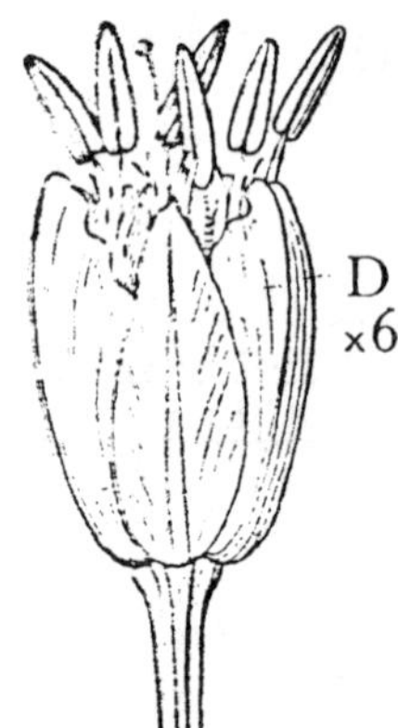


A) inner perianth segment and anther of <u>A. vineale</u>
B) perianth segment and anther of <u>A. sphaerocephalon</u> (not in key)
C) outer perianth segment and anther of <u>A. vineale</u>
D) exserted anthers, e.g. <u>A. vineale</u>, <u>A. carinatum</u>.

## *JUNCUS COMPRESSUS* Jacq./*J. GERARDI* Lois.

*Juncus compressus* is very scarce in the region, though perhaps overlooked. It superficially resembles *J. gerardi* Lois. but differs in a combination of variable characters. It is a plant of calcareous, wet meadows and flushes.

Rhizome usually 5cm, rarely far-creeping; flower stems curved, not stiffly erect, compressed throughout their length; inflorescence normally shorter than the lowest bract, capsule obtuse, about $1\frac{1}{2}$ x as long as the light-brown perianth, anthers slightly shorter than the filaments; style shorter than the capsule. Not in salt marshes. *J. compressus* Jacq.

Rhizome far-creeping, plant forming more extensive tufts or patches; flower stems stiffly erect, compressed below, triquetrous above, inflorescence usually considerably exceeding the lowest bract, capsule acute, rarely exceeding the dark-brown to blackish perianth, anthers 3 x as long as the filaments, style at least as long as the capsule. Salt marshes.
*J. gerardi* Lois.

## *JUNCUS ALPINUS* Vill./*J. ARTICULATUS* L./*J. ACUTIFLORUS* Ehrh.

These three species, though sometimes superficially similar, are well separated on several characters. *J. alpinus* occurs only in Teesdale (in the region) growing in marshy places, gravelly streamsides, etc.

1. Perianth segments obtuse, the outer mucronate, the inner not mucronate; inflorescence with few branches arising from two points about 4cm apart; capsule obtuse, apiculate or mucronate.
*J. alpinus* Vill. (*J. alpinoarticulatus* Chaix.)

1. Perianth segments acute; inflorescence usually much-branched; capsule acute or acuminate.

2. Capsule long-ovoid, contracted above to an acumen; outer perianth segments acute, the inner more blunt with broad colourless margins; leaves strongly laterally compressed, usually curved; plant with branched rhizome, and therefore often subcaespitose in habit.
*J. articulatus* L.

2. Capsule evenly tapered to an acute point; perianth segments acute, the outer tapering to outcurved awnlike points, the inner with brownish margins; leaves terete or slightly compressed; rhizome far-creeping with closely-spaced nodes. *J. acutiflorus* L.

The habit of *J. acutiflorus* can be likened to that of tightly-packed *Carex arenaria*, i.e. a linear arrangement of shoots rather than subcaespitose.

## *JUNCUS* x *SURREJANUS* Druce (*J. acutiflorus* Ehrh. x *J. articulatus* L.)

Variable in habit, though ± intermediate between the parents. The best way to detect a hybrid is by its non-swelling capsule and absence of seed. Hybrids are best detected in the field, where they can be compared with the particular variant of *J. articulatus* growing nearby.
Probably widespread, but rarely reported.

*J.* x *buchenaui* Dorfl. (*J. alpinus* x *J. articulatus*) has been recorded in v.c. 66.

## JUNCUS BUFONIUS agg.

Five species within the Juncus bufonius aggregate are now recognised in Europe, three of which occur in Britain and in our area. Of the three, only J. bufonius is widespread and common. J. foliosus has a western distribution, occurring in freshwater habitats such as on the muddy margins of ponds, lakes, streams, and rivers, in wet fields, roadside ditches, and on waste land. J. ambiguus is typically a halophyte, occurring on the coast on mud and salt-flats above high water mark, and inland on salt-flashes and salt-flats, and lime-waste tips.

1. Leaves bright green, more than 1.5mm wide; tepals usually with a dark line on either side of the midrib; anthers 1.2-5.0 times as long as filaments; seeds with 20-30 conspicuous longitudinal ridges (use a x20 hand-lens). J. foliosus Desf.

1. Not with the above combination of characters. Leaves usually darker and seldom more than 1.5mm wide; seeds apparently smooth or with a minutely reticulate surface.

2. Inflorescence partly (rarely wholly) contracted; inner tepals obtuse or rounded, often emarginate and mucronate at tip; capsule truncate, as long or longer than inner tepals. J. ambiguus Guss.

2. Inflorescence with widely spaced flowers (rarely partly or wholly contracted); inner tepals usually acute, sometimes sub-acute; capsule acute or sub-acute, rarely truncate, and then clearly shorter than inner tepals. J. bufonius L.

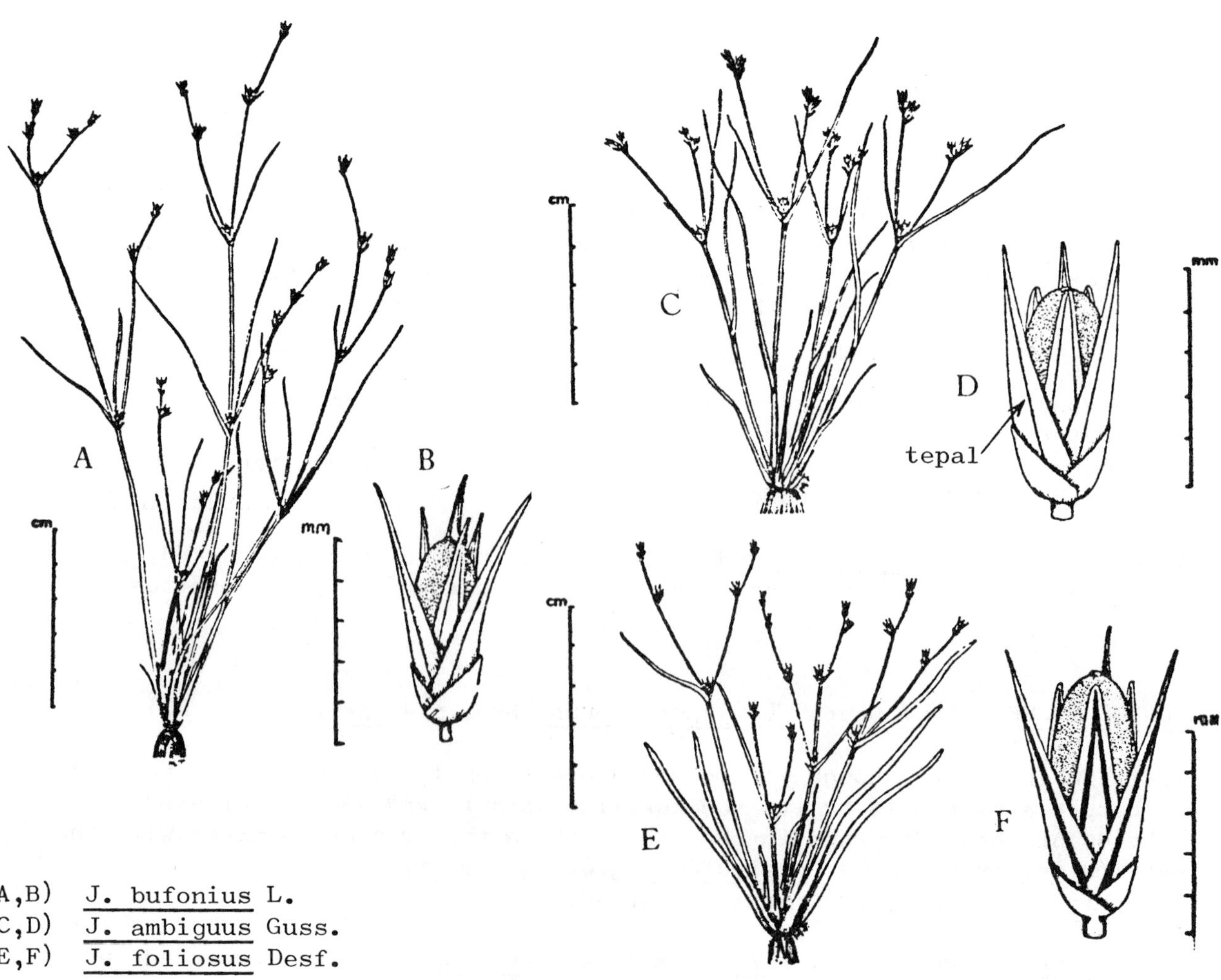


A,B) J. bufonius L.
C,D) J. ambiguus Guss.
E,F) J. foliosus Desf.

T.A. Cope & C.A. Stace, Watsonia 12 113-128 (1978)

JUNCUS x DIFFUSUS Hoppe (J. effusus L. x J. inflexus L.)

The hybrid resembles J. inflexus in inflorescence shape, but the floral characters, stem striation and anatomy are intermediate. The stem has 18-40 striae below the inflorescence. The capsule is smaller than in either parent, and seed production is much reduced (but very variable).
J. x diffusus has been detected in very few places in the region, and should be looked for wherever the parental species are growing together.

Stace, (1975).

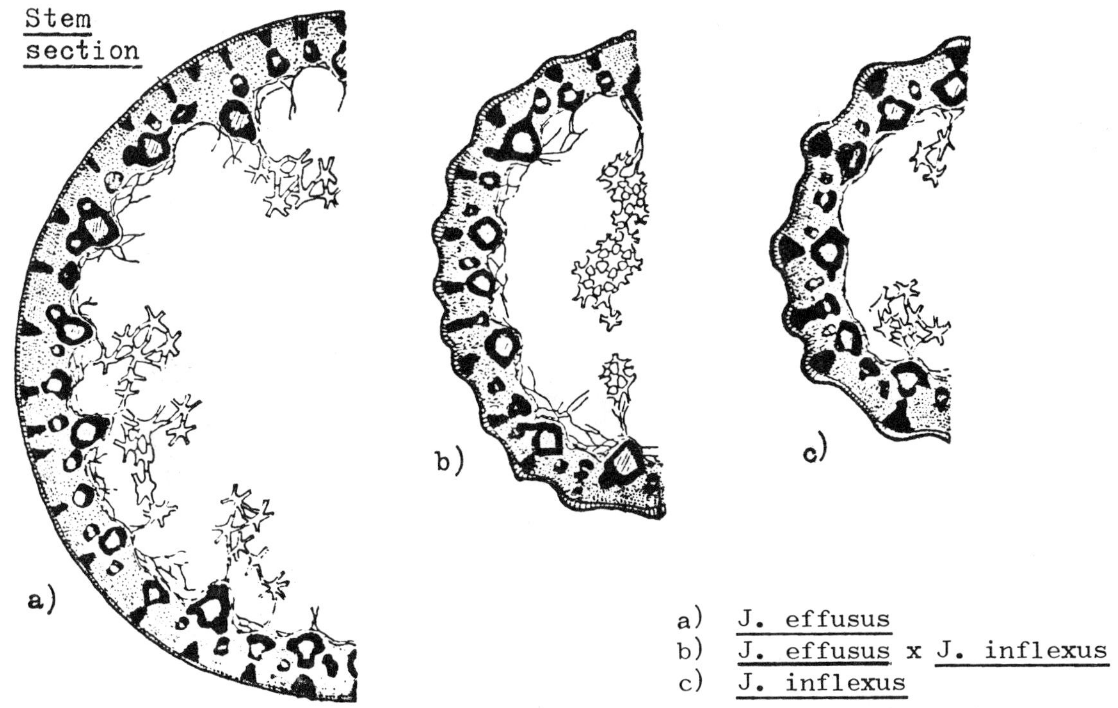


a) J. effusus
b) J. effusus x J. inflexus
c) J. inflexus

O. Nilsson & S. Snogerup, Bot. Notiser 124 179-186 (1971) - drawings

JUNCUS BULBOSUS L.

There is no consensus as to whether the two species J. kochii and J. bulbosus are distinct, or as to the characters used to distinguish them. Most recorders have not separated them, so very little information on distribution is available in the region. A key to the species is given below, so that recorders can test the distinctness or otherwise of these taxa.

Stamens 3 (rarely 6); anthers elongate, about equalling the filaments; plant less robust, with leaves 3-10cm long. J. bulbosus L.

Stamens 6; anthers shortly-oblong, distinctly shorter than the filaments; plant more robust with leaves 8-12cm long. J. kochii F.W. Schultz

P. Benoit & D.E. Allen, Proc. B.S.B.I. 7(3) 504 (1968)

LUZULA CAMPESTRIS (L.)DC./L. MULTIFLORA (Retz.)Lej.

These species are easily confused, and overlap in many characters. L. multiflora is mainly confined to acid soils, but L. campestris is more catholic in its requirements, occurring on both acidic and basic soils. The shape of the seeds is diagnostic and can easily be seen by splitting a ripe capsule. Note that the pale appendage often remains attached to the seed, thus apparently altering the length/breadth ratio. Note also that green unripe seeds of L. campestris may not be swollen, thus appearing similar to those of L. multiflora. Stolons and anthers also afford good distinguishing characters. Plant size is not necessarily of value.

Usually loosely tufted, with short stock and shortly-creeping stolons. Anthers 2-6 x as long as the filaments. Seeds nearly globose, with a white basal appendage up to half their length. L. campestris(L.)DC.

Usually densely tufted with few or no stolons. Anthers about as long as filaments. Seeds oblong, nearly 2x as long as broad, with a white basal appendage up to half their length. Probably over-recorded for L. campestris. L. multiflora(Retz.)Lej.

x12

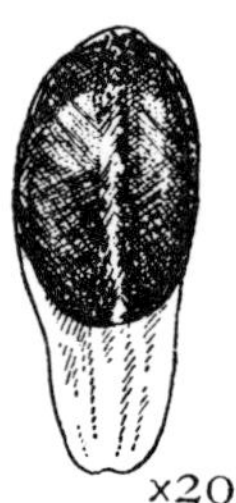

x20

Seeds of L. campestris L. multiflora

IRIS VERSICOLOR L.

Iris versicolor L. is naturalised at Ullswater and Esthwaite, and the hybrid I. versicolor x I. virginica occurs on the shores of Windermere. They both have purplish flowers. The hybrid is morphologically similar to I. versicolor and is appreciably, but not completely sterile, setting 0-2 capsules per flowering shoot. The capsules have 0-12 seeds which often have degenerated ovaries.

J R Ellis, in Stace (1975).

EPIPACTIS Sw.

Six of the seven British species of Epipactis are found in the region, E. purpurata being the exception. All except E. helleborine are local and/or rare. E. atrorubens is locally frequent in limestone areas, especially on limestone pavement in Westmorland. Several colonies of E. leptochila have recently been discovered in an unusual habitat (river gravels) by the Tyne in v.c.67. Species are rather variable in height, leaf-shape, pubescence, etc, but floral characters are more stable. To identify the species, careful examination of the flowers is necessary.

Cross-polination can be ascertained by the presence of a rostellum in the freshly opened flower, which when touched with, for example, the point of a pencil, will draw out the pollinia. Cross-pollination is also indicated by the absence of pollinia from the majority of older flowers. However, in exposed situations the rostellum dries up soon after the flower opens, and self-fertilisation may ensue.

In self-fertilised species the rostellum is not visible, the pollinia cannot be removed with a pencil (or by insects), and the pollen eventually crumbles and fertilises the stigma of the same flower.

1. Rhizome long; flowers brownish outside; hypochile with 2 erect, lateral lobes; epichile connected by a narrow hinge. Fen and fen-pasture.
E. palustris (L.)Crantz

1. Rhizome short; flowers greenish or purple outside; hypochile without lateral lobes; epichile connected by 1 or more folds. Not in fen or fen-pasture.

2. Bracts subglabrous; flowers subglabrous, hanging vertically, entirely apple-green except for labellum, often cleistogamous; self-fertilised; perianth persistent in fruit. E. phyllanthes G.E. Sm.

2. Bracts usually obviously pubescent; flowers ± patent; perianth purplish or green, shrivelling after fertilisation, not cleistogamous.

3. Cross-fertilised. Flowers purplish or green.

4. Leaves small, lanceolate (largest 3-8cm), length:breadth ratio 1:3-4, leaves and stem suffused with violet; flowers yellowish-green, inside of hypochile whitish-violet, often speckled. E. purpurata Sm.

4. Leaves lanceolate to ovate, often large (largest (3)6-15cm), length: breadth ratio 1:1.5-3.0, dark green, rarely violet suffused; flowers dull green or purple.

5. Ovaries noticeably pubescent; rachis with thick, crisped pubescence; internode under lowest flower much longer than the others; flowers usually red-purple; epichile with large crinkly bosses; hypochile red inside. Plant often very small. E. atrorubens (Hoffm.)Schult.

5. Ovaries subglabrous; pubescence of rachis not thick or crisped; internode under lowest flower not or scarcely longer than the others; flowers dusky-purple to apple-green; epichile with small, smooth bosses; hypochile dark brown or dark green inside. Plants often very large. E. helleborine (L.)All.

3. Self-fertilised. Flowers pale green except for pale pink labellum.

6. Plant slender; flowers small; epichile about as long as broad.
E. dunensis (Steph.)Godf.

6. Plant robust; flowers large; epichile longer than broad.
E. leptochila (Steph.)Godf.

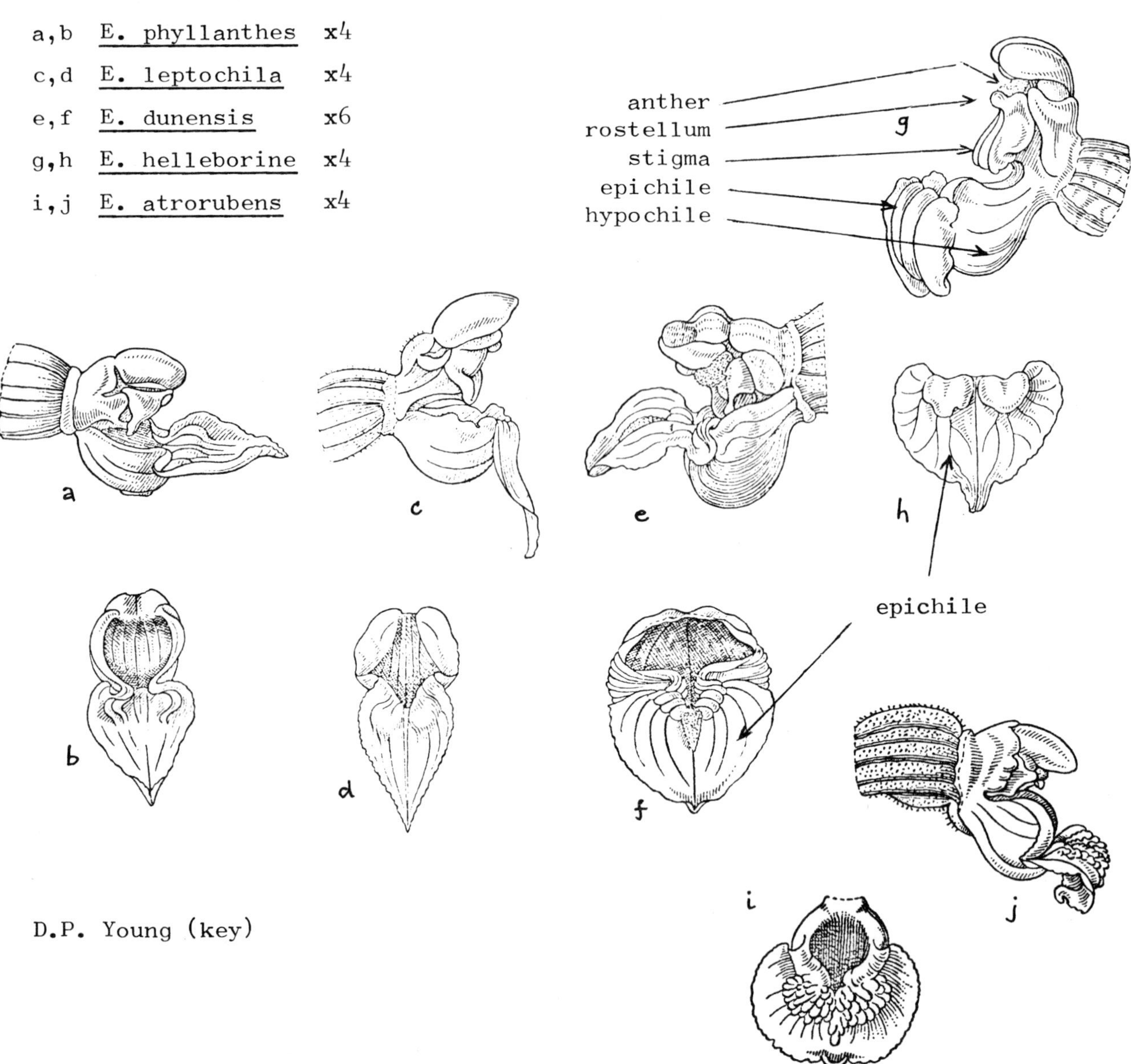


D.P. Young (key)

GYMNADENIA CONOPSEA (L.)R.Br.
Subsp. DENSIFLORA (Wahlenb.) G. Camus, Bergon & A. Camus

This taxon is a striking plant, normally much taller than the common subspecies, with dense spikes more than 10 cm long, and leaves more than 15 mm wide. The flowers are bright rose-red or magenta with the smell of cloves.

It is a plant typically of fens and marshes, with a few records in the region.

DACTYLORHIZA Nevski x GYMNADENIA R.Br

Three intergeneric hybrids have been reported in the region, but not in the present recording period. Plants appearing to have characters of both Dactylorhiza and Gymnadenia should be sought wherever species of both genera grow in close proximity.

x Dactylogymnadenia cookei (H-Harrison f.)Soó
(D. fuchsii x Gymnadenia conopsea)
x D. varia (T. & T.A. Steph.)Soó (D. purpurella x G. conopsea)
x D. legrandiana (Camus)Soó (D. maculata x G. conopsea)

DACTYLORHIZA Nevski

Problems of identification within this genus arise both from the variability of the species, and from the wide occurrence of hybrids. Careful assessment of all characters is essential. Vegetative characters are more variable than the floral.

Hybridisation and backcrossing give rise to populations intermediate between the parent species, or populations showing more or less complete intergradation between the parents. Further complexity arises from intergeneric crosses involving, in particular, Coeloglossum and Gymnadenia. Where hybrids are suspected, a careful examination of the population should be carried out. Note in the field, all the following characters in all plants in the population:

| | |
|---|---|
| leaf colour | marks on labellum |
| hooding of leaves | shape and colour of labellum |
| width of leaves | habitat |
| hollowness of stem | |

1. Lateral sepals spreading widely or drooping; flowers white, pale or medium-pink, or pale magenta-pink, or lilac - with darker lines, loops, or dots on labellum and petals; stem solid; leaves usually spotted; spur slender (2 mm or less in diameter); labellum broad, 3-lobed in lower half.

2. Middle lobe of labellum longer than the side lobes, or all three equal in length; lowest leaves broadest, and very blunt or rounded at apex, usually progressively narrower and more acute, total number 7-20, usually with numerous round or elliptic purplish-black spots which are often wider at right-angles to length of leaf, silver-grey on under-surface; spur 1.5-2 mm. in diameter.
D. maculata (L.) Soó subsp. fuchsii (Druce) Hyl.
(D. fuchsii (Druce) Soó, Orchis fuchsii Druce)

2. Middle lobe of labellum much smaller and usually shorter than lateral lobes; all leaves rather narrow and acute, total number 5-12 spots on leaves often rather faint or obscure, not broader at right-angles to length of leaf; spur 1 mm or less in diameter. D. maculata (L.) Soó subsp. ericetorum (E.F. Linton) Hunt & Summerh.
(Orchis ericetorum E.F. Linton)

1. Lateral sepals ± upright, backs often touching; flowers flesh-pink, madder-red, pale-magenta, magenta-purple - with darker markings on labellum or petals, or pale creamy straw-coloured; stem often with centre hollow; leaves spotted or unspotted; spur rather stout, over 2 mm in diameter, often ± conical.

3. Leaves unspotted.

4. Leaves erect, yellow-green, narrowly-hooded at apex; stem with wide hollow; labellum not lobed, red and purple pigment present, usually with distinct double-loop markings, sides strongly reflexed; spur stout, tapering towards apex. D. incarnata (L.)Soó (Orchis strictifolia)

a) Labellum flesh-pink. Base-rich areas. D. incarnata subsp. incarnata

b) Labellum madder-red. Mainly dune-slacks.
D. incarnata subsp. coccinea

c) Labellum magenta or purple. Acidic heaths and bogs.
D. incarnata subsp. pulchella

4. Leaves broad, usually more than 2cm wide, numerous (usually 5 or more), spreading, deep- or greyish-green, not or broadly hooded at apex; stem solid or narrowly-hollow; labellum scarcely lobed, the sides usually not much reflexed, usually broader than long.

5. Labellum large, with dots or indistinct loops, nearly all in the centre; flowers lilac-pink, magenta-pink, or pale magenta; leaves large, lanceolate, erect, fresh- or deep-green. Occurs at Teesmouth.
D. majalis (Rchb.)Verm. subsp. praetermissa
(Druce) Moresby-Moore & Soó
(D. praetermissa (Druce) Soó)

5. Labellum not or only slightly equally 3-lobed, often broadly diamond-shaped, and magenta or bright magenta with heavy broken-loop, dash or hieroglyphic markings; leaves broadly hooded at apex, plants up to 45cm high. D. majalis (Rchb.)Verm. subsp. purpurella
(T. & T.A. Steph.) Moresby-Moore & Soó
(D. purpurella (T. & T.A. Steph.) Soó)

3. Leaves variously spotted.

6. Spots on leaves very small, dot-like, usually towards apex; flowers bright-magenta with heavy broken line markings on labellum; labellum broadly diamond-shaped; leaves broadly hooded at apex. D. majalis subsp. purpurella

Several hybrids have been found in the region.

| | |
|---|---|
| D. x transiens (Druce)Soó | (D. fuchsii x D. maculata) |
| D. x kernerorum (Soó)Soó | (D. fuchsii x D. incarnata) |
| D. x venusta (T. & T.A. Steph.)Soó | (D. fuchsii x D. purpurella) |
| D. x grandis (Druce)P.F. Hunt | (D. fuchsii x D. praetermissa) |
| D. x claudiopolitana (Soó)Soó | (D. incarnata x D. maculata) |
| D. x formosa (T. & T.A. Steph.)Soó | (D. maculata x D. purpurella) |
| D. x wintoni (A. Camus)P.F. Hunt | (D. incarnata x D. praetermissa) |
| D. x latirella (P.M. Hall)Soó | (D. incarnata x D. purpurella) |

## SPARGANIUM L.

The four British species of Sparganium occur in the region. S. minimum and S. angustifolium are widespread, but local, and found mainly in Cumbria.

To determine the subspecies of S. erectum, fully ripe fruit must be present. Subsp. microcarpum is the common subspecies, but subsp. neglectum occurs locally in Northumberland, and perhaps elsewhere.

1. Inflorescence branched; male capitula on lateral branches; perianth segments black-tipped. *S. erectum* L. (*S. ramosum* Huds.)

1. Inflorescence not branched; male capitula on main axis; perianth segments not black-tipped.

2. Cauline leaves keeled and sheathing at base, but not inflated, ± triangular in section; male capitula usually more than 3, distant. *S. emersum* Rehm. (*S. simplex* Huds.)

2. Cauline leaves not keeled at base, flat in section; male capitula usually 1-2, crowded.

3. Leaf-like bract of lower female capitulum 10-60cm long, at least twice as long as whole infloresence; leaves very long, sheaths somewhat inflated at base; male capitula 1-3, and ± contiguous on main axis; lower female capitula peduncled. Fruit about 8mm long. *S. angustifolium* Michx.

3. Leaf-like bract of lower female capitulum 1-5(8)cm long, barely exceeding the inflorescence; leaves usually 2-6mm wide, barely sheathing at base; male capitulum usually 1 (rarely 2); female capitula (1)2-3, usually sessile. Fruit 3.5-4.5mm. *S. minimum* Wallr.

fruit of *S. erectum* subsp. *microcarpum*

fruit of *S. erectum* subsp. *neglectum*

SCIRPUS LACUSTRIS L.
(Schoenoplectus lacustris (L.)Palla)

The two subspecies which are now recognised within this taxon, were formerly each given species status. Whilst subsp. tabernaemontani is for the most part obviously glaucous and subsp. lacustris not, this distinction does not always hold, and the species should always be confirmed by examining the floral parts.

Stems not glaucous; glumes smooth (awn often papillose); stigmas usually 3, nut at least bluntly trigonous. subsp. lacustris

Stems glaucous; glumes densely beset with small, dark-brown, bluntly-hooked papillae on back, especially near the midrib; stigmas 2; nut biconvex, or plano-convex. subsp. tabernaemontani (C.C. Gmel.) Syme
(S. tabernaemontani (C.C. Gmel.)Palla)

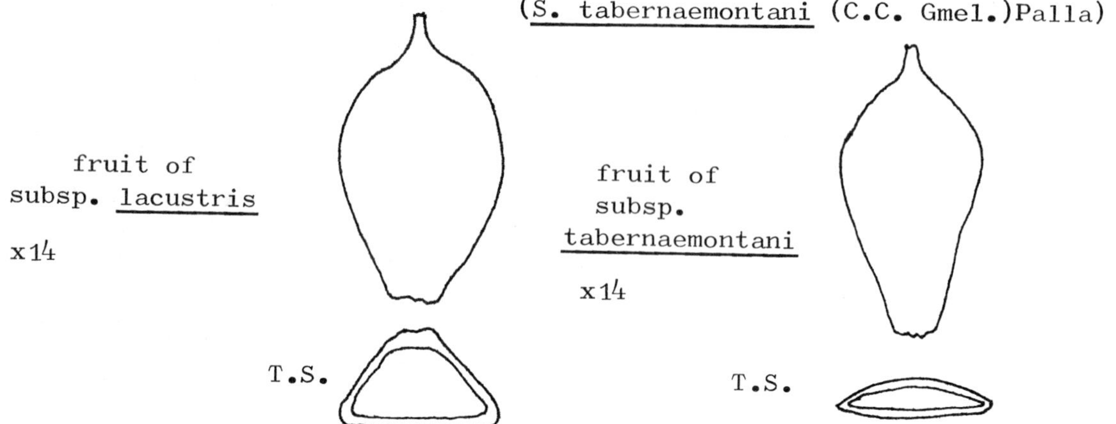


SCIRPUS CERNUUS Vahl/S.SETACEUS L.
(Isolepis cernua (Vahl) Roem. & Schult./I. setacea (L.) R.Br.)

In S. setaceus the bract is occasionally shorter than the inflorescence. Without reference to other distinguishing characters, such plants can easily be mis-identified as S. cernuus.

Bract usually distinctly longer than inflorescence; spikelets usually 2-3; nut longitudinally ribbed, shiny. S. setaceus L.

Bract shorter than or slightly exceeding inflorescence; spikelet usually solitary; nut smooth, not shiny. S. cernuus Vahl

ELEOCHARIS R. Br.

All British species of Eleocharis except E. parvula are found in the region. They generally present few problems of identification, except sometimes in the E. palustris/uniglumis group.

Note that the stems of E. acicularis are not always clearly 4-angled.

1. Lowest glume at least half as long as spikelet.

2. Glumes about 2mm long; rhizome creeping, brown; stems usually 4-angled; spikelet 3-4mm. Wet sandy or muddy places, sometimes submerged, and then sterile. E. acicularis (L.) Roem. & Schult.

2. Glumes about 5mm long; rhizome short, stout, producing slender runners; stems never 4-angled; spikelet 5-7mm. Damp peaty places, usually base-enriched. E. quinqueflora (F.X. Hartmann) Schwarz
(E. pauciflora (Lightf.) Link)

1. Lowest glume much less than half as long as spikelet.

3. Plant densely tufted; upper sheath very obliquely truncate and acute; stigmas 3; nut trigonous. E. multicaulis (Sm.) Sm.

3. Plant not densely tufted; upper sheath almost transversely truncate; stigmas 2; nut biconvex.

4. Lowest glume not more than half encircling base of spikelet.

5. Bristles 4(very rarely 0); style base broader, markedly constricted at junction with the nut; spikelet cylindrical.
E. palustris (L.) Roem. & Schult.

5. Bristles (4)5(6); style base narrower, hardly constricted at junction with nut; spikelet narrowly-conical, pointed.
E. austriaca Hayek

4. Lowest glume ± completely encircling base of spikelet.
E. uniglumis (Link)Schult.

E. austriaca is known at a few sites in Yorkshire, Cumberland, and Northumberland, and thorough search may yet lead to its discovery in Durham. It usually grows on a gravel substrate with some silt deposition where there is slow water movement and some protection from spates. Typically in pools, runnels, and springs beside upland rivers, but seldom along the banks of the main river. Apart from the fruiting characters, it differs from E. palustris in:

i) the absence or poor development of reddish-purple colour at the base of the stem.

ii) the spikes characteristically compact and narrowly-conical and pointed, in contrast with the cylindrical spikes of E. palustris.

iii) stems somewhat inflated and brittle, with a circular cross-section.

iv) extremely high proportion of fertile stems borne rather close together on the relatively short rhizome.

E. uniglumis sometimes presents difficulties. The typical plant with thin stems in salt-marshes is easy to identify, especially if the stem is well-developed to show:

a. the single glume encircling the spikelet,

b. the fertile second glume, and

c. the punctate pattern of the ripe nut.

However, there are puzzling populations of plants intermediate between E. palustris and E. uniglumis. Some of these have been shown to have intermediate chromosome numbers and are probably of hybrid origin. Furthermore, plants with a high polyploid number (2n = c.90) may occur in mixed populations with the ordinary E. uniglumis (2n = 46).

/ drawings overleaf

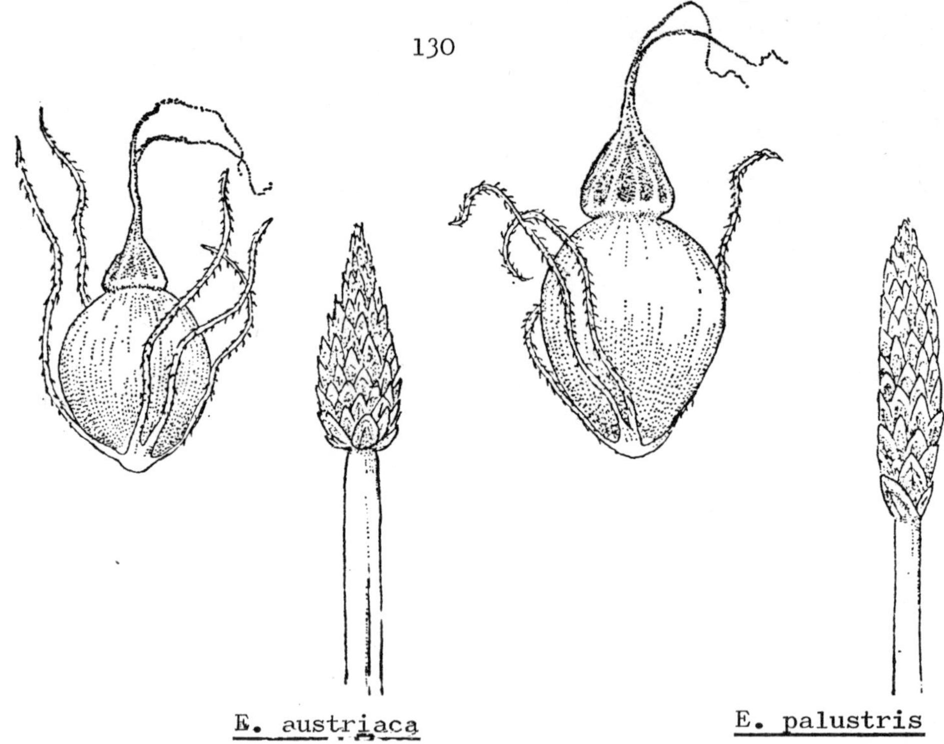

E. austriaca E. palustris

CARREX L.

For the identification of Carex species, 'British Sedges' by A.C. Jermy & T.G. Tutin is essential. The notes below supplement or merely emphasise the information given in that publication.

CAREX DISTANS L./C. PUNCTATA Gaud./C. HOSTIANA DC./C. BINERVIS Sm.

These four species are closely related and can be similar in general appearance. They can be separated on a combination of characters.

C. hostiana is distinct in the trigonous apex of the leaf, and in the contrasting bright yellow-green utricle and dark-brown glume. It usually has rather yellower foliage than the others.

C. punctata is characterised by the wide, sharply-bifid beak of the utricle, by the wider leaves, and less reliably by other features (R.W. David, pers. comm. 1979). It has not been found in the region.

C. distans and C. binervis can be confused especially if growing in lowland/coastal habitats. They can be separated as follows:

| | C. distans | C. binervis |
|---|---|---|
| leaves | grey-green, becoming brown, then ash-grey and persisting on dying. | dark-green, with wine-red blotches on dying, persisting as pink- or orange-brown litter. |
| sheaths | dark- or mid-brown, becoming fibrous | red-brown, not becoming fibrous |

| | | |
|---|---|---|
| ripe utricles | green, rarely dark-brown lateral ribs present, but not as prominent | purple-brown, rarely green, with 2 prominent lateral ribs |
| ♀ glume | brown, rarely chestnut | dark purple-brown |
| nut | yellow-brown, ellipsoid | olive-brown, flattened above |

The lateral ribs of the ripe utricles are not as diagnostic as illustrations infer. Mature utricles of C. distans sometimes have strong lateral ribs, but with the other ribs also pronounced.

A.C. Jermy & T.G. Tutin, 1968.

CAREX FLAVA Group

Taxa in this group are, in general, not well-defined, and further revision is needed. The overlapping ranges of variation in most or all of the characters make the demarcation of taxa difficult. When the taxa grow in close proximity, hybrids may occur, but not commonly.

The ratio of stem-length : leaf-length, and the arcuate utricle of C. lepidocarpa is usually sufficient to separate it from C. demissa. The latter species comprises a complex of forms, some of which are sometimes mistakenly identified as C. lepidocarpa. Hybrids between these two species are probably overlooked, but those specimens appearing intermediate are usually forms of C. demissa.

1. Ripe utricle 6.0-6.5mm, gradually tapering into a beak 2.0-2.5mm long; leaves 4-7mm wide, about as long as stem; ligule 5mm. C. flava L.

1. Ripe utricle 2.0-5.0mm, with distinct shoulder at base of beak, or more gradually tapering; leaves 2-4(5)mm wide, $\frac{1}{4}$ to 1 times as long as stem; ligule c. 1mm.

   2. Ripe utricle yellowish-green or -brown, 3.5-5.0mm, arcuate-deflexed, with distinct shoulder; beak 1.5-2.0mm, $\frac{1}{3}$ or more the length of utricle; leaves $\frac{1}{4}$ to $\frac{1}{2}$ length of stem; stems straight. C. lepidocarpa Tausch

   2. Ripe utricle usually green, 3.0-4.0mm, straight, gradually narrowed into the straight beak; beak c. 1.0mm, less than $\frac{1}{3}$ length of utricle; leaves 1.5-5.0mm wide, often $\pm$ as long as stem; stem often curved. C. demissa Hornem.

   2. Ripe utricle 2.0-3.5mm, straight; beak 0.5-1.0mm, not or only slightly deflexed; leaves 1.5-3.0mm wide, $\pm$ as long as stem, flat or channelled, or even inrolled. C. serotina Mérat

The hybrid C. flava x C. demissa is known in v.c.69. Plants show much reduced pollen fertility, and backcrossing to both parents appears to occur.

C. serotina is usually distinct in its small, crowded utricles.

A.C. Jermy & T.G. Tutin, 1968, and A.O. Chater, pers.comm. (1980)

CARREX ACUTIFORMIS Ehrh./C. RIPARIA Curtis

Purely vegetative material of these two species presents considerable problems of identification. Because of the large overlap in the ranges of variation of many characters said to be useful in separating them, only those specimens outside the range of overlap can be named with confidence.

| | C. riparia | C. acutiformis |
|---|---|---|
| i) leaves | erect, glaucous, 6-15mm wide, tapering rather abruptly in upper 1/5 of leaf | arcuate, glaucous at first, 7-10mm wide, tapering gradually in upper $\frac{1}{3}$ of leaf |
| ii) inner sheaths | translucent, with distinct transverse septa | not translucent, with less distinct transverse septa |
| iii) ligule | 5-10mm, obtuse or rounded | 5-15mm, usually acute |
| iv) fibrillae | rarely formed | usually fibrillose |

The tapering of the leaf is often difficult to determine, and sometimes the tapering begins in the lower $\frac{1}{2}$ of the leaf. Character ii) is also difficult to interpret, and often appears unworkable. The ligule of C. acutiformis is sometimes very short and obtuse.

The leaf transverse section is said to be a good differentiating character (N.T.H. Holmes, pers. comm. 1979).

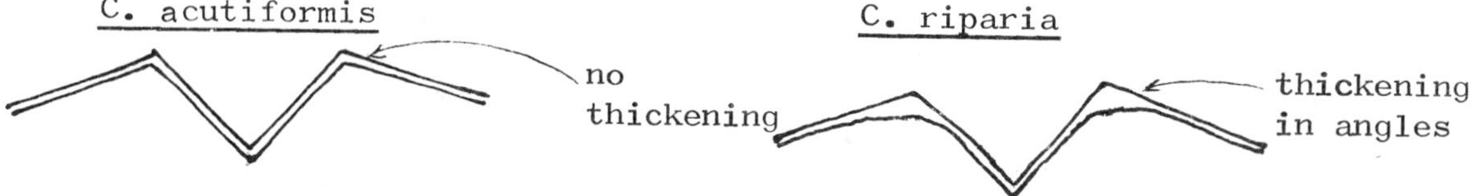


This character was investigated (in fruiting plants) in 1979, but it was not found to be consistent. Further investigation is needed.

It is not clear to what extent hybridisation occurs between these two species. Stace, 1975, states that the hybrid remains unconfirmed in Britain, although it has been widely reported.

CAREX LIMOSA L./C. MAGELLANICA Lam. (C. PAUPERCULA Michx.)

These species are similar in general appearance and habitat, and in the region occur mainly in Cumbria. C. limosa grows at the edges of pools or in very wet blanket- or valley-bog, or in mesotrophic mires. C. magellanica (C. paupercula) is a plant of wet Sphagnum-bogs (blanket- or valley-bog).

| | C. limosa | C. magellanica |
|---|---|---|
| leaves | shorter than stems | often equalling stems |
| lowest bracts | not exceeding inflorescence | exceeding inflorescence |
| ♀ spikes | without ♂ flowers | with ♂ flowers at extreme base |
| ♀ glumes | 3.5-4.5mm, slightly wider than ripe utricle, ovate, apex acute | 5.0-6.5mm, narrower than ripe utricle, lanceolate, apex acuminate or aristate |
| ripe utricle | strongly ribbed, beak c. 0.5mm | faintly ribbed,beak 0. |

A.C. Jermy & T.G. Tutin, 1968

CAREX CARYOPHYLLEA Latour./C. ERICETORUM Poll.

These two species are very similar, but are separated as follows:

| | C. caryophyllea | C. ericetorum |
|---|---|---|
| stems | trigonous | bluntly-trigonous |
| leaves | 1.5-2.5mm wide | to 4mm wide |
| ligules | 1.0-2.0mm long | less than 1.0mm long |
| ♂ glumes | 4-5mm, red-brown, apex acute | 2.5-3.0mm, purple-brown, apex rounded, margins scarious, fringed. |
| ♀ glumes | red-brown, midrib excurrent | purple-brown, midrib not excurrent, margin scarious, fringed. |
| ripe utricle | ovoid-ellipsoid, green | subglobose, green below, dark-brown above |

C. ericetorum is rare, but is now known at several sites in Westmorland and one in Durham. C. caryophyllea is widespread and common.

A.C. Jermy & T.G. Tutin, 1968

C. ORNITHOPODA Willd./C. DIGITATA L.

C. digitata is similar to C. ornithopoda in habit and general structure, but is usually larger in all its parts with more widely-spaced spikelets.

| | C. ornithopoda | C. digitata |
|---|---|---|
| leaves | deeper green, 5-20cm, glabrous | lighter green, up to 25cm, usually sparsely-hairy on upper surface |
| inflorescence | 1/8 - 1/10 length of stem, compact, ± digitate | 1/5 - 1/4 length of stem, |
| ♂ glumes | c. 2.5mm, obovate | 5mm, rounded or even emarginate at apex |
| ♀ spikes | 2-3, 5-10mm, 2-4 flowered (rarely more), ± sessile | 1-2, 10-20mm, 5-10 flowered, pedunculate |
| ♀ glumes | 2.0-2.5mm | 3.0-4.0mm |
| ripe utricles | 3.0-4.0mm, much exceeding glumes | 3.0-4.0mm, less tapered below, about equalling glumes |

When flowering or fruiting these species should be easily distinguishable, for the lowest spike of the inflorescence of C. digitata is clearly separated from the one above it, whereas in C. ornithopoda all the spikes originate from almost the same point; the inflorescence of C. digitata is sometimes at least tinged with crimson while that of C. ornithopoda is straw-coloured; and in C. digitata the female glumes are as long as the utricles whereas in C. ornithopoda they are markedly shorter. When the plants are in the vegetative state separation is not so easy, for the intensity of the red colouration of the basal sheaths and the breadth and degree of hairiness of the leaves (the distinctions usually quoted) are relative. However, the sheaths of C. digitata are, in general, more deeply and genuinely crimson (as opposed to mid- or dark-green). Furthermore, the new shoots of C. digitata are very distinctive. They begin to appear in October and are then tinged with deep red and tipped with green. In March they elongate and arch over at the tips, presenting a highly characteristic fountain-shape. None of these characters is found in C. ornithopoda.

C. ornithopoda is a plant mainly of upland limestone scars, and C. digitata of lowland limestone, often lightly-wooded. Their distribution overlaps in a limited area of S. Westmorland.

A.C. Jermy & T.G. Tutin, 1968.
R.W. David, Watsonia 13(1) 53-54 (1980)

CAREX NIGRA Group

Of the species in this group, C. nigra shows the greatest variation in morphology in response to habitat conditions. For example, stems 7-70cm, ♂ spikes 5-30mm, ♀ spikes 7-50mm, leaves short and wide to long and narrow. Species in the group may be separated as follows:-

1. Plant densely-tufted or forming tussocks, with long pioneer rhizomes; leaf-sheaths splitting to form fibrillae; lowest bract much shorter than inflorescence. *C. elata* All.

1. Plants always with pioneer rhizomes, although poorly-developed in stagnant water; leaf-sheaths membranous, not forming fibrillae; lowest bract as long as or longer than inflorescence.

2. Leaves 1.5-3.0mm wide; lowest bract equalling or slightly exceeding inflorescence; ♀ spikes 7-50mm. Plant very variable. *C. nigra* (L.) Reich.

2. Leaves 3-7mm wide; lowest bract much exceeding inflorescence; ♀ spikes 20-100mm long.

3. Leaves bright-green and shiny beneath; lower sheaths vinous-red-brown; stem brittle, usually breaking when folded, smooth; ♀ glumes appressed to fruit. *C. aquatilis* Wahlenb.

3. Leaves dull beneath; lower sheaths not or only faintly tinged with red; stem not breaking when folded, rough above; ♀ glumes stiff, often giving the spikelet a prickly appearance. *C. acuta* L.

Of the species in this group, only *C. nigra* is common and widespread in the region. *C. elata* and *C. acuta* are local, and *C. aquatilis* distinctly rare. *C. aquatilis* is found by rivers in northern Britain, and is probably under recorded in the region.

CAREX MURICATA L. Complex

1. Utricles with nut well above the corky swollen base, and with narrow elongated serrate beak, the whole 4.5mm or more long; glumes acuminate, ± tawny, giving, with the long-beaked utricles, a shaggy look in the inflorescence; leaves and stems slightly fleshy, ligule soft, narrow, 4-8 (-10)mm long; roots, and sometimes (but not invariably) basal sheaths, ligules, and glumes tinged vinous-purple. *C. spicata* Huds. (*C.contigua* Hoppe, *C.muricata* auct. *non* L.)

1. Not as above.

2. Inflorescence not more than 3cm long, spikes ± contiguous; ligule neatly ovate, slightly longer than broad; base of utricle flat or rounded with the nut set close upon it.

3. Flowering stems erect, rigid; glumes markedly shorter than utricles, dark, contrasting in colour with the utricles until these ripen and darken; utricles rounded, with a broad margin or flange, narrowing abruptly into a short beak; strongly calcicole, flowering May. *C.muricata* L. subsp. *muricata* (*C.pairaei* F.W.Schultz subsp. *borealis* Hylander)

3. Flowering stems flexuous; glumes ± as long as utricles, yellow or pale brown, concolorous with unripe utricles but becoming white and then contrasting with the brown ripe utricles; utricles ovoid, narrowly margined, narrowing evenly into the beak; calcifuge, flowering June. *C.muricata* L. subsp. *lamprocarpa* Čelak. (*C.pairaei* F.W.Schultz, *C.bullockiana* Nelmes)

2. Inflorescence distinctly separated, with intervals of 1-3cm between the lowest two spikes; ligule as long as or shorter than broad; utricle narrowed at base as well as at beak i.e. diamond-shaped.

4. Plant flaccid, dark- or grey-green, leaves often as long as the drooping flowering-stem; inflorescence very interrupted, up to 10cm long or even more with intervals of 2cm or more between the lowest spikes; utricles 3.5-4.0(4.5)mm, appressed to the stem-axis, becoming greyish black when fully mature. C.divulsa Stokes subsp. divulsa

4. Plant robust, erect, bright yellow-green, leaves shorter than flowering stem; inflorescence not more than 6cm long, with intervals of less than 2cm between lowest spikes; utricles (4.0)4.5-4.8mm long, markedly divaricate, becoming red-brown when mature.
C.divulsa Stokes subsp. leersii (Kneucker) W.Koch
(C.leersii F.W.Schultz non Willd., C.polyphylla Kar. & Kir.)

The above descriptions are of the extreme forms of the two subspecies of C. divulsa, which are very distinct; but many intermediates occur, at least in southern Britain.

R.W. David, 1979.

GLYCERIA R.Br.

All the taxa occur in the region, though G.declinata and G. x pedicellata are often overlooked. G.maxima is distinctly rare.

1. Ligules broad, with median tooth 3-6mm long; plants 90-250cm high, reed-like, with stout to robust culms; panicles much-branched; lemmas 3-4mm long. G. maxima (Hartm.) Holmberg

1. Ligules blunt or pointed, without a median tooth; plants 10-100cm high, not reed-like; panicles scarcely or moderately-branched; lemmas 3.5-7.5mm long.

2. Lemmas 6.0-7.5mm long, acute to subobtuse; anthers 2-3mm; sheaths smooth. G. fluitans (L.) R.Br.

2. Lemmas 3.5-5.5mm long; anthers less than 2mm; sheaths smooth or rough.

3. Lemmas 4.0-5.0mm long, sharply-toothed at tip; anthers 0.8(1.0)mm; leaves greyish-green, abruptly-contracted at apex. G. declinata Bréb.

3. Lemmas 3.5-5.5mm long, not sharply-toothed at tip, but sometimes with a distinct 'shoulder'; anthers 1.0-1.8mm; leaves not greyish-green.

4. Spikelets caducous, soon disarticulating; lemmas 3.5-5.0mm long, very blunt, rounded or 'shouldered'; anthers 1.0-1.5mm, dehiscing, with viable pollen; sheaths rough or minutely-hairy near blade.
G. plicata Fries

4. Spikelets persistent; lemmas (4)5.0-5.5mm long, intermediate in shape between the parents; anthers 1.0-1.8mm, not dehiscing, with inviable pollen; sheaths often minutely rough near blade.
G. x pedicellata Towns.
(G. fluitans x G. plicata)

The usually widely-spreading inflorescence branches of G.plicata are easily seen at a distance. By contrast, the branches of G. fluitans remain at a very narrow angle, looking almost appressed for most of the time. The hybrid G. x pedicellata is intermediate in this character.

The hybrid can be recognised readily in late summer by the intact spikelets, when those of G. fluitans and G. plicata have already disarticulated.

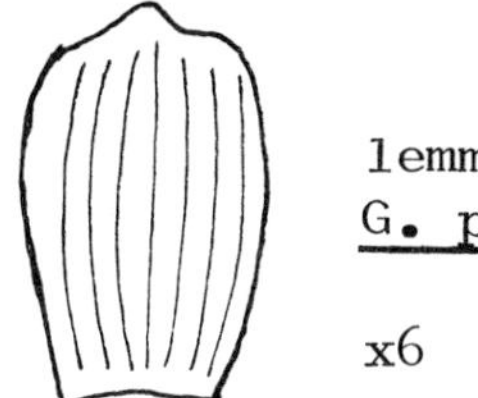

lemma of
G. plicata

x6

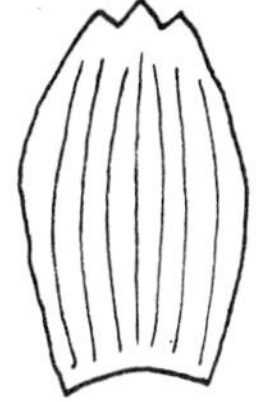

lemma of
G. declinata

x6

FESTUCA RUBRA L./F. OVINA L. complex

This group presents difficulties to some observers mainly because of the small or subtle differences between the species. It is important to examine the rooting system to determine whether a plant is rhizomatous or tufted, and the branching, which may be intra- or extravaginal.

1. Spikelets viviparous (proliferous). F. vivipara (L.) Sm.

1. Spikelets not viviparous.

2. Plants loosely or densely matted, or with scattered shoots and culms; rhizome present, very slender, extensively-creeping; branching extravaginal. F. rubra L. and F. juncifolia St.-Amans (see below)

2. Plant densely-tufted, without rhizomes; branching intravaginal (except sometimes in F.rubra subsp. commutata).

3. Leaf-blades of two kinds - the basal very fine 0.3-0.5mm wide, infolded, those of the culm flat, 2.0-4.0mm wide. F. heterophylla Lam.

3. Leaf-blades of one kind, those at the base similar to those of the culm.

4. Lemmas finely-pointed, usually awnless, 2.5-3.5mm long; leaf-blades hair-like, 0.2-0.4mm in diameter; spikelets 3-7mm long. Fairly common in acidic grassland. F. tenuifolia Sibth.

4. Lemmas 3.5-8.0mm long, tipped with a fine awn up to 6mm long; leaf-blades usually wider; spikelets 5-12mm long. Habitats various.

5. Leaf-blades 0.3-0.5mm wide; leaf-sheaths with free margins; lemmas 3-5mm long, tipped with an awn 0.5-1.5mm long. F. ovina L.

5. Leaf-blades 0.5-1.0mm wide; leaf-sheaths tubular or with free-margins; lemmas 4.0-5.5mm long, tipped with an awn to 4mm long.

6. Leaf-sheaths tubular, with margins united, but soon splitting; branching sometimes extravaginal. Widely used in seed-mixtures for lawns. F. rubra L. subsp. commutata Gaud.

6. Leaf-sheaths with free margins. F. longifolia Thuill.

F. longifolia and F. rubra subsp. commutata are very easily overlooked, and are perhaps sometimes recorded erroneously as F. rubra subsp. rubra. Both may occur in the region.

F. ovina and F. tenuifolia are distinguished vegetatively from F. rubra subsp. rubra by their tufted habit, intravaginal branching, sheaths with free margins, and lack of rhizomes. F. tenuifolia has been much overlooked, and recorded as F. ovina. It should be noted that F. ovina is not common in lowland areas (except on limestone), and tends to be mistakenly recorded as F. rubra (the latter species has extravaginal branching, long, thin, far-creeping rhizomes, and tubular leaf-sheaths).

Extra- and Intravaginal branching

This is a good character for distinguishing F. rubra from the F. ovina complex. If the leaf-sheaths are stripped off layer by layer, one can see in the intravaginal species that the young shoots do not break through the lower sheaths, whereas in the extravaginal species, the young shoots can be seen breaking through.

Tuftedness is liable to misinterpretation. It is, therefore, important to confirm the presence or absence of rhizomes by carefully digging up one or two plants.

## F. RUBRA L./F. JUNCIFOLIA St.-Amans

F. juncifolia closely resembles some maritime varieties of F. rubra subsp. rubra, especially the var. arenaria which also grows on sand-dunes and has a similar loose form of growth. One can, however, define a taxon identifiable with F. juncifolia by means of a combination of the following characters:

i) long upper and lower glumes and lemmas
ii) rigid, inrolled leaf-blades, with well-developed and sometimes continuous abaxial sclerenchyma.
iii) extensively-creeping extravaginal, but no intravaginal branches.
iv) octoploid chromosome number, 2n = 56.

| | Lower glume | Upper glume | Lemma | Lemma-awn |
|---|---|---|---|---|
| F. juncifolia | 4.2 - 6.7 mm | 5.5 - 8.6 mm | 6.5 - 8.9 mm | 0.1 - 2.1 mm |
| F. rubra | 2.2 - 4.1 mm | 3.0 - 5.8+ mm | 4.4 - 7.0+ mm | 0.1 - 3.3 mm |

C.A. Stace, and R. Cotton, Watsonia, 10, 119-138 (1974)

## X FESTULOLIUM Aschers & Graebn. (Festuca L. x Lolium L.)

Hybrids of Festuca pratensis Huds. and Lolium spp. resemble Lolium perenne L. in having spikelets arranged alternately in two rows on opposite sides of the usually unbranched axis. Hybrids are very variable, probably because of backcrossing.

The spikelets resemble those of F. pratensis, but are borne on very short pedicels (or sometimes pedicels absent), and have two glumes, the lower one much reduced or rarely absent.

X Festulolium loliaceum (Huds.) P.Fourn (F.pratensis Huds. x Lolium perenne L.) has unawned lemmas.

X Festulolium braunii (K. Richt.) A. Camus (F.pratensis x Lolium multiflorum Lam.) has awned lemmas.

POA PRATENSIS L. and allies

Hubbard's Grasses is essential. The following key and notes supplement the information given there.

1. Tillers forming compact tufts. Vegetative tillers with very narrow, bristle-like leaves, usually less than 2mm wide. Lemmas 2-3mm long, rarely more. P.angustifolia L.

1. Tillers in tufts or singly-borne. Vegetative tillers broad, and more than 2mm wide. Lemmas 3-5mm long.

2. Plants with tillers normally in tufts, but sometimes not clearly tufted; panicle-branches normally 3-5 at lowest rachis node; lower glume normally 1-nerved; lemmas 3-4mm. P. pratensis L.

2. Plants with single tillers, arising from slender rhizomes, not tufted; panicle-branches 1 or more at lowest rachis node; lower glume normally 3-nerved; lemmas 2.5-5mm.

3. Culms normally 4-6 noded; panicle stiff usually narrow, but sometimes more open; panicle branches 1 or more; spikelets small, blunt, not 'heavy' in appearance; leaf-blade with no hairs at base (i.e. by the ligule); glumes 2-3mm; lemmas 2.5-3mm conspicuously blunt in side view. P. compressa L.

3. Culms normally 1-2 noded; panicle loose and open; panicle-branches at lowest node usually 2, sometimes 1 or 3+; spikelets fairly large, pointed, and 'heavy' in appearance; leaf-blade with a conspicuous fringe of hairs at base (i.e. by the ligule); glumes 3-5mm; lemmas 3-5mm, pointed in side view. P. subcaerulea Sm. (P. pratensis subsp. irrigata (Lindm.) Lindb.f.)

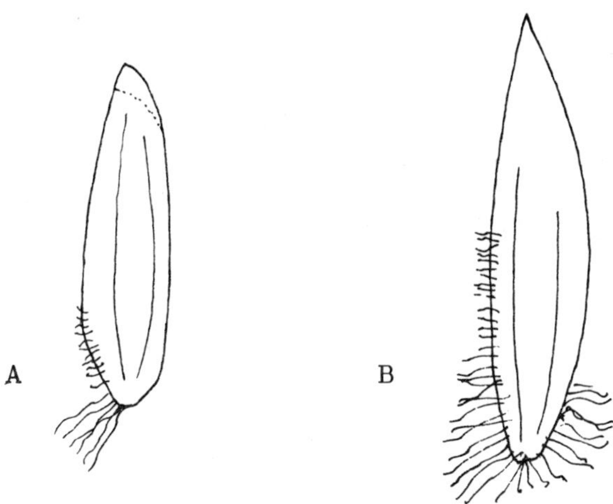


Side view of Lemma of A: P. compressa
B: P. subcaerulea — both x6

P. subcaerulea is much overlooked, and is sometimes misidentified as P. compressa. Both species can have a flattened stem.

BROMUS L.

Certain taxa in this genus are sometimes very similar, and identification depends upon close examination of specimens. Most are uncommon or rare, but, with the exception of B. pseudosecalinus, all have been recorded in the region, though some not recently.

1. Grain thick, longitudinally inrolled ; margins of lemma inrolled, not or scarcely overlapping in fruit ; rachilla disarticulating tardily ; panicle lax.

2. Lemma not more than 6.5mm long ; lower sheaths softly-hairy.

3. Lemma 6-6.5mm ; anthers (3)4-5mm ; panicle very lax, patent or drooping ; stems 80-110cm long. B. arvensis L.

3. Lemma 5-6mm ; anthers less than 3mm ; panicle narrow, contracted ; stems not more than 60cm. B. pseudosecalinus P.M.Sm.

2. Lemma 6.5-9(10)mm ; anthers 1-2mm ; lower sheaths glabrous or very sparsely hairy. B. secalinus L.

1. Grain thin, flat or feebly-rolled ; margins of lemma gaping and overlapping in fruit ; rachilla fragile, soon disarticulating ; panicle lax or dense.

4. Lemma 4.5-6.5mm, with wide, hyaline, sharply-angled margin ; ripe grain exceeding palea. B. lepidus Holmberg

4. Lemma usually more than 6.5mm, with narrow, hyaline, rounded or bluntly-angled margin ; ripe grain not exceeding palea.

5. Panicle usually dense, most pedicels shorter than their spikelets ; lemma papery, with prominent veins ; anthers rarely more than 1mm , usually less.

6. Stems 1-8(12)cm long, prostrate or procumbent to ascending ; lemma 6.5-7.5mm ; awns sometimes divaricate in fruit ; grain shorter than palea. B. hordeaceus L. subsp.thominii (Hard.)Maire & Weiller

6. Stems 3-80cm long, usually erect ; lemma 6.5-11mm ; awns straight, erect ; grain equalling or shorter than palea.

7. Lemma 6.5-8mm, usually glabrous, with broad or narrow, normally bluntly-angled hyaline margins ; awn 3-7mm ; grain usually equalling palea. B. x pseudothominii P.M.Sm. (B. hordeaceus subsp.hordeaceus x B. lepidus)

7. Lemma 8-11mm ,usually hairy, with narrow, bluntly-angled hyaline margin ; awn 4-11mm ; grain shorter than palea. B. hordeaceus subsp. hordeaceus

5. Panicle lax, most pedicels longer than their spikelets ; lemma corneous, with obscure veins ; anthers more than 1mm.

8. Anthers (3)4-5mm ; panicle open and very loose, up to 30cm ; palea about equalling lemma. B. arvensis L.

8. Anthers 0.2-3mm ; panicle loose or ± compact, not more than 15cm ; palea shorter than lemma.

9. Lemmas cucullate, 6.5-8mm long, with rounded margins, anthers 1.5-3mm ; rachilla internode 0.5-1mm ; panicle narrow, usually erect. B. racemosus L.

9. Lemmas not cucullate, 8-11mm long ; with bluntly-angled margins ; anthers to 1.5mm ; rachilla internode about 1.5mm ; panicle broad (rarely narrow), erect at first, eventually drooping. B. commutatus L.

P.M. Smith, in Flora Europaea, Vol. 5 (1980)
P.M. Smith, Watsonia 9 319-332 (1973)
P.M. Smith, Watsonia 6 327-344 (1968)

B. commutatus and B. racemosus are similar morphologically and are often confused by recorders. B. racemosus has a cucullate lemma which is narrower in relation to its length than is that of B. commutatus. The markedly convex back of the lemma of B. racemosus seen from the side (Fig.F), lends a distinctive appearance to the spikelets in which the upper part of each lemma appears to be slightly bulged outwards below its hyaline tip and margin. The lemma-margin in bluntly-angled in B. commutatus and gradually rounded in B. racemosus (Fig. A, B). Intermediate conditions occur, however.

B. racemosus is often a plant of waysides, and is therefore liable to be mistakenly identified as B. hordeaceus. Close examination of populations of Bromus in such habitats may reveal the presence of B. racemosus.

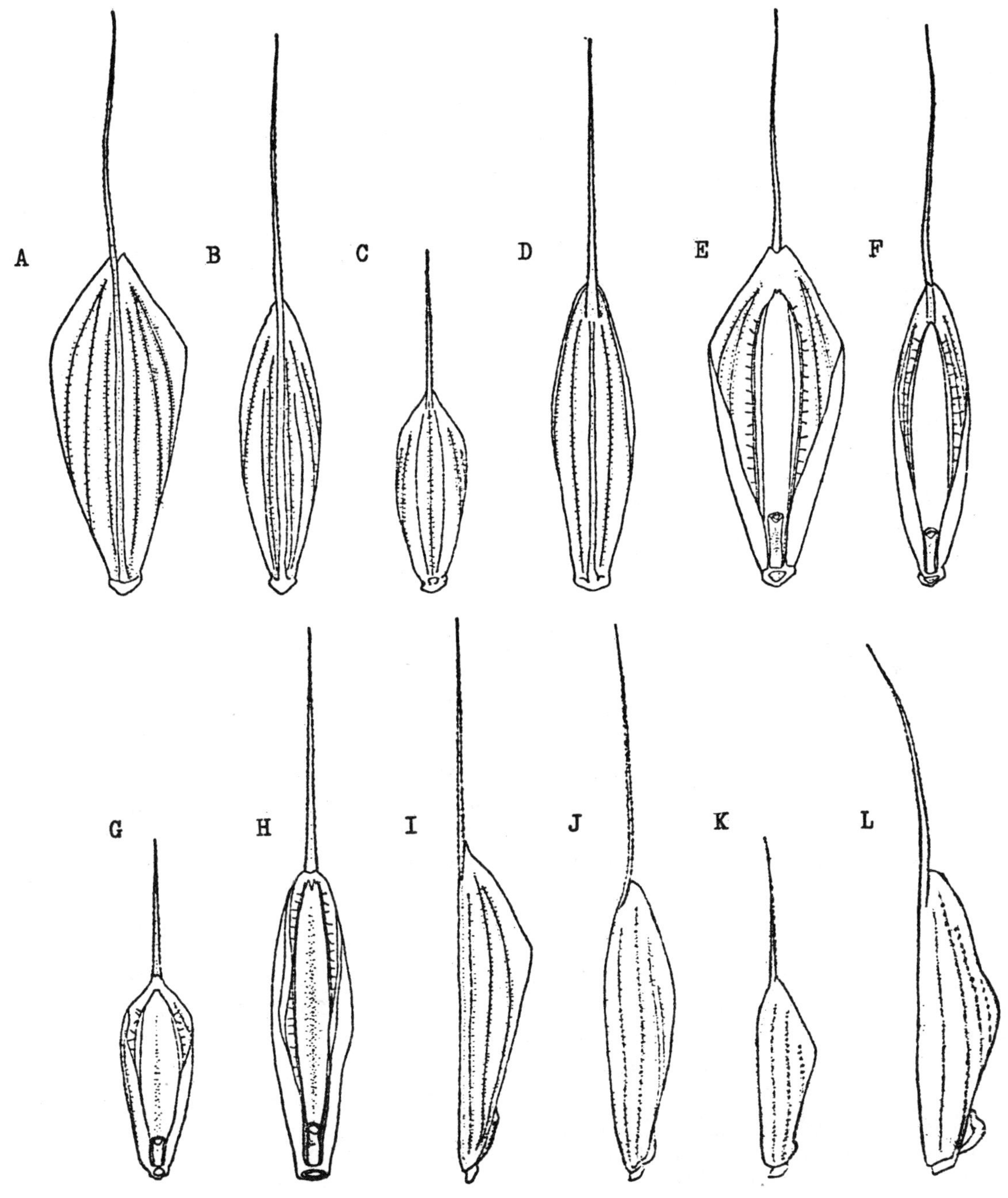


Abaxial (A-D), Adaxial (E-H), and side views (I-L) of basal florets
B. commutatus (A,E,I) B. racemosus (B,F,J) B. pseudosecalinus (C,G,K)
B. secalinus (D,H,L)

ELYMUS L. (AGROPYRON Gaertn.)

A recent revision has seen the Britsh species of Agropyron sensu lato placed in Elymus L. The coastal taxa of Elymus (Agropyron) have presented problems because of their variability and the not infrequent occurrence of hybrids. Note that Elymus arenarius sensu lato is now renamed Leymus arenarius (L.) Hochst.

1. Rachis fragile, disarticulating at maturity, glabrous on main angles ; lemmas 11-20mm long ; leaf-sheaths without auricles. Maritime sands.
E. farctus (Viv.)Runemark ex Melderis subsp. boreali-atlanticus (Simonet & Guinochet)Melderis (A. junceiforme (A. & D. Löve) A. & D. Löve)

1. Rachis tough, not disarticulating at maturity, spinose-ciliate on the main angles; lemma 8-15 mm long; leaf sheaths with auricles (sometimes very small).

2. Leaves flat or tightly-convolute, with prominent, closely-crowded,flattened veins, usually spinose-ciliate on main angles; lemma 8-15 mm long.

3. Spikelets closely overlapping, 1/5 to 1/2 their length apart, 10-20 mm long; pollen viable.
E. pycnanthus (Godron) Melderis (A. pungens (Pers) Roem. & Schult)

3. Spikelets 1/2 their length or more apart, 15-28 mm long; spikes loose; pollen shrivelled, inviable. Sterile hybrids between E. farctus or E. pycnanthus and E. repens.

4. Leaf-blades minutely-rough on the closely- and prominently-ribbed upper surface.
E. x obtusiusculum (Lange) (E. pycnanthus x E. farctus)

4. Leaf-blades with short-haired, less-ribbed upper surface.
E. x laxum (Fr.) (E. farctus x E. repens)

2. Leaves usually flat, with fine or prominent, not closely-crowded, rounded, often sparsely-hairy veins on upper surface; lower sheaths usually not ciliate; rachis sometimes shortly-hairy.
E. repens (L.) Gould subsp. repens (A. repens (L.) Beauv.)

E. repens var. arenosus (Petif.) Melderis (A. maritimum (Koch & Ziz.) Jansen & Watcher) is an obscure taxon which could key out under E. pycnanthus. It is said to differ in its smooth ribs and its small, distant spikelets. Of possible occurrence on sand dunes.

A. Melderis, in Flora Europaea, Vol 5 (1980)
A.J. Silverside, pers.comm, 1977
T.G. Tutin, pers. comm, 1978

AVENULA (Dumort.) Dumort. (HELICTOTRICHON Besser)

Leaves stiff, entirely glabrous; lower sheath glabrous; panicle with 1-2 branches at lowest rachis node; spikelets 3-6 flowered.

Avenula pratense (L.) Dumort.
(Helictotrichon pratense (L.) Pilger)

Leaves softer, usually hairy; lower sheath hairy; panicle with 3-4 branches at lowest rachis node; spikelets 2-3 flowered.

Avenula pubescens (Huds.) Dumort.
(Helictotrichon pubescens (Huds.) Pilger)

The lower sheaths of A. pubescens are sometimes very sparsely-hairy, but plants can always be separated from A. pratense in floral and other characters.

HOLCUS L.

When vegetative, the two species can be separated readily be examination of the rooting system. H. lanatus is invariably tufted and not rhizomatous, whilst H. mollis always has long, creeping rhizomes. Hairyness is variable in both species. H. lanatus also varies considerably in colour and is sometimes not purple-striped at base. In H. mollis the awn sometimes projects very little beyond the glume, such plants then liable to be mistaken for H. lanatus. It is, therefore, always best to confirm the identity by determing whether the plant is tufted or rhizomatous.

AGROSTIS L.

Species of Agrostis have small, 1-flowered spikelets arranged in a much-branched, loose to contracted panicle. Calamagrostis species differ in having longer spikelets with numerous silky hairs from the base, some of which are at least half as long as the lemma.

For certain determination of the species, the whole plant should be used including rootstock, vegetative shoots and inflorescence.

1. Palea rudimentary and apparently absent (x10 lens); lemmas usually awned from the lower half.

2. Plant spreading by slender, trailing stolons, often forming a close turf; rhizomes absent. Plant strongly calcifuge, wet places.
A. canina L. subsp. canina

2. Plant tufted, spreading by short, slender rhizomes usually less than 0.7mm in diameter bearing scale-leaves 1.0-1.7mm wide. Drier habitats on heath and moorland. A. vinealis Schreber
(A. canina L. subsp. montana (Hartm.) Hartm.)

1. Palea $\frac{1}{2}$ to $\frac{3}{4}$ as long as lemma; lemma usually awnless.

3. Plant spreading by stolons which frequently root at the nodes; rhizomes absent; panicle usually contracted after flowering. A. stolonifera L.

3. Plant rhizomatous; panicle usually remaining open after flowering.

4. Rhizomes abundant and conspicuous, stout (at least some exceeding 2mm in diam.), bearing scale-leaves 5-7mm wide when opened out. Disturbed habitats, arable land, etc. Rare. *A. gigantea* Roth.

4. Rhizomes short, less freely produced, not exceeding 1.5mm diam., bearing scale-leaves about 2.5mm wide. In permanent grassland. *A. capillaris* L. (*A. tenuis* Sibth.)

Hybrids have been reported in Britain between all pairs of species (except between *A. canina* and *A. gigantea*, and are only certainly recognised by possessing abortive pollen.

The presence in the same floret of an awned lemma and a conspicuous palea would suggest a hybrid between *A. canina*, and *A. stolonifera* or *A. capillaris* (*A. tenuis*).

*A. gigantea* Roth. is usually larger and coarser than *A. capillaris* but it has the same open panicle, spreading in both flower and fruit. It differs from the latter species also in the longer ligules to the sterile shoots. Sometimes the curved or procumbent culm-bases of *A. gigantea* root at the nodes, and then there can be confusion with *A. stolonifera* L. However, in *A. gigantea* tough, creeping rhizomes are always present. In *A. stolonifera* these rhizomes are absent, but extensive stolons are often formed, giving rise to a mat or loose turf. In the latter species, the panicle closes after flowering, becoming dense and almost lobed.

If the stolons of *A. stolonifera* are buried, they become blanched and are easily mistaken for rhizomes. Because of this, great care should be taken with specimens from arable fields where harrowing or hoeing has taken place.

*A. canina* L. nearly always has awns and a minute palea, distinguishing it from all other species (except *A. curtisii* (*A. setacea*), which does not occur in the region). It isw a plant of acidic grasslands, common in the uplands, but more local in lowland areas.

The presence or absence of a conspicuous palea is sometimes difficult to determine in the field. If specimens are gathered at anthesis and dried, the spikelets gape and observation of the palea is facilitated.

## *PHLEUM PRATENSE* L./*P. HUBBARDII* D. Kov. (*P. BERTOLONII* DC.)

These species are distinct in several characters; nevertheless, small specimens of *P. pratense* are still being misidentified as *P. hubbardii*. Note that although *P. pratense* is normally the taller, more robust species, there is considerable overlap in culm-length and panicle-length.

| | *P. hubbardii* (*P. bertolonii*) | *P. pratense* |
|---|---|---|
| Culm-length (cm) | 3-8 | 7-15 |
| Leaf length (cm) | 5-12(15) | 12-30 |
| " width (mm) | 2-3 | 6-9 |
| Ligule-length (mm) | 2-4 | 5-6 |
| Panicle-length (mm) | 10-80(90) | (30)60-150 |
| " width (mm) | 3-6 | (6) 7-10 |
| Spikelet length (mm) | 2.0-2.5 | 3.0-3.5 |
| Awn-length (mm) | 0.5-1.0 | 1.0-2.0 |

P. hubbardii is indicated when the width of the middle leaf plus the width of the panicle is less than 10mm. However, the above measurements must be checked for confirmation.

D. Kováts, Acta Bot. Acad. Sci. Hung. 23, 119-142, (1977)

EUPHORBIA ESULA L. / E.URALENSIS Fisch. ex Link / E. CYPARISSIAS L.

There is much dispute over the limits of the taxa in this critical group. E. cyparissias is the most distinct in having linear stem-leaves, about 2mm wide. The other two species have been differently interpreted by different workers. Moore (1958) considers that E. esula and E. uralensis (E. virgata Waldst. ex Kit., E. esula subsp. tommasiniana (Bertol.)Nyman) are best separated on leaf characters :

E. esula - leaves linear-lanceolate or obovate-linear, broader above the middle, tapering to the base, and sessile.

E. uralensis - leaves lanceolate to ovate, broader below the middle, some leaves widening at the base, with a short petiole.

Hybrids are intermediate in character, but are difficult to recognise, and have been confused with both parents, and with hybrids between them and E. cyparissias. There is no agreement as to whether the common plant in Britain is E. esula, or E. uralensis, or the hybrid between them. Because of this, specimens should be collected for later determination.

P.M. Benoit and C.A. Stace, in Stace, 1975
R.J. Moore, Canad. J. Bot. 36 547-559 (1958)

## OCCURRENCE OF SPECIES IN VICE-COUNTIES 66 - 70

The list below includes those taxa, mentioned in the main text, which lack a recent record from at least one of the five vice-counties. The remaining taxa mentioned in the main text have been recorded recently in all five vice-counties, unless stated to be absent from the region.

| taxon | vice-county | 66 | 67 | 68 | 69 | 70 |
|---|---|---|---|---|---|---|
| Isoetes setacea | | | | | | + |
| I. lacustris | | | | | + | + |
| Equisetum x litorale (E. arvense x E. fluviatile) | | + | | | + | + |
| Ophioglossum azoricum | | | | +) | | + |
| Hymenophyllum tunbrigense | | | | + | | + |
| H. wilsonii | | | + | | + | + |
| Polypodium australe | | | | | + | |
| P. x font-queri (P. australe x P. vulgare) | | | ? | | | |
| P. x mantoniae (P. interjectum x P. vulgare) | | | | | + | + |
| Asplenium billotii | | | | | | + |
| A. trichomanes / trichomanes | | | + | | | + |
| A. x murbeckii (A.ruta-muraria x A.septentrionale) | | | | | | + |
| A. x alternifolium (A. septentrionale x A. trichomanes / trichomanes) | | | | + | | + |
| Polystichum x bicknellii (P.aculeatum x P.setiferum) | | | | | | +) |
| Dryopteris x tavelii (D. filix-mas x D. pseudomas) | | + | | | + | |
| D. x mantoniae (D. filix-mas x D. oreades) | | | | | + | |
| D. expansa | | + | + | | + | + |
| D. x brathaica (D. carthusiana x D. filix-mas) | | | | | + | |
| D. x ambrosiae (D. austriaca x D. expansa) | | | | | | + |
| D. x deweveri (D. carthusiana x D. austriaca) | | | | | + | |
| D. villarii / submontana | | | | | + | +) |
| Juniperus communis / nana | | | +) | | + | + |
| Ranunculus circinatus | | +) | | +) | +) | +) |
| R. baudotii | | + | +) | + | +) | |
| R. reptans | | | | | + | + |
| R. x bachii (R. aquatilis x R. fluitans) | | | + | + | | |
| Papaver lecoqii | | + | + | + | + | |
| Fumaria capreolata | | +) | | + | + | + |
| F. purpurea | | +) | | +) | + | + |
| F. bastardii | | + | | +) | + | + |
| F. officinalis / wirtgenii | | + | | + | | +) |
| Barbarea stricta | | +) | | +) | + | + |
| B. verna | | +) | | + | + | +) |

| taxon | 66 | 67 | 68 | 69 | 70 |
|---|---|---|---|---|---|
| Barbarea intermedia | + | +) | + | +) | + |
| Raphanus maritimus | +) | | +) | + | + |
| Lepidium campestre | + | + | +) | + | + |
| L. perfoliatum | | +) | | | +) |
| L. latifolium | + | +) | + | + | +) |
| L. sativum | + | +) | + | + | + |
| L. ruderale | +) | +) | | +) | +) |
| Erophila verna / spathulata | + | + | + | +) | + |
| Cardamine impatiens | | +) | | + | + |
| Rorippa amphibia | + | | | | |
| R. x sterilis (R. nasturtium-aquaticum x R. microphyllum) | + | + | + | +) | + |
| Viola rupestris | + | | | + | |
| V. reichenbachiana | + | +) | +) | + | + |
| V. canina | +) | + | + | + | + |
| V. x burnatii (V. riviniana x V. rupestris) | + | | | | |
| V. reichenbachiana x V. riviniana | + | | | | |
| V. tricolor / curtisii | | +) | +) | + | + |
| V. lutea x V. tricolor | | +) | | | |
| V. arvensis x V. tricolor | +) | | | | |
| Hypericum x desetangsii (H. perforatum x H. maculatum) | + | | | | + |
| Elatine hexandra | | | | + | + |
| Silene noctiflora | + | + | + | + | +) |
| Cerastium tomentosum | ?+ | | + | | |
| C. biebersteinii | + | | | | |
| Stellaria neglecta | + | +) | + | + | + |
| Sagina apetala / apetala | | + | + | + | + |
| Minuartia verna | + | + | +) | + | + |
| Arenaria serpyllifolia /leptoclados | + | + | + | +) | |
| Montia fontana / fontana | + | + | + | +) | +) |
| M. fontana / chondrosperma | + | + | + | +) | |
| M. fontana / amporitana | + | + | | +) | +) |
| M. fontana / variabilis | + | + | + | +) | +) |
| Chenopodium polyspermum | + | | +) | +) | +) |
| C. glaucum | + | +) | +) | | |
| C. ficifolium | + | +) | +) | | |
| C. vulvaria | + | +) | | | |
| C. murale | + | +) | +) | | |
| Atriplex glabriuscula | + | | | + | + |
| A. longipes | | | | + | |
| A. longipes x A. prostrata | + | | | | |
| Salicornia europaea | | + | + | + | + |

| taxon | vice-county | 66 | 67 | 68 | 69 | 70 |
|---|---|---|---|---|---|---|
| Salicornia ramosissima | | + | + | | +) | + |
| S. fragilis | | | | | + | |
| S. dolichostachya | | + | + | + | | + |
| Geranium rotundifolium | | +) | +) | | | |
| Erodium moschatum | | + | | | | |
| E. cicutarium / bipinnatum | | | | | + | + |
| E. maritimum | | +) | | | +) | +) |
| Oxalis exilis | | + | | | | |
| O. corniculata | | + | | +) | + | + |
| O. stricta | | | | +) | | + |
| O. europaea | | | | | + | +) |
| O. incarnata | | + | | | | +) |
| Ononis spinosa | | + | | | + | + |
| Trifolium micranthum | | + | | + | +) | +) |
| Anthyllis vulneraria / lapponica | | | + | + | | |
| Vicia angustifolia s.s. | | + | | | + | + |
| Potentilla x mixta (P. erecta or P. anglica x P. reptans) | | + | | | + | + |
| P. x suberecta (P. erecta x P. anglica) | | + | | | + | + |
| Agrimonia procera x A. eupatoria | | | +) | | | |
| Alchemilla monticola | | + | | | | |
| A. acutiloba | | + | + | | | |
| A. mollis | | | + | + | | + |
| A. gracilis | | | + | | | |
| A. filicaulis / filicaulis | | + | + | + | + | |
| A. subcrenata | | + | | | | |
| A. glomerulans | | + | | | + | |
| A. wichurae | | + | | | + | + |
| Rosa arvensis | | + | + | + | + | + |
| R. obtusifolia | | ?+ | | +) | | |
| R. rubiginosa | | + | + | + | + | +) |
| R. micrantha | | +) | | | | |
| R. tomentosa | | + | + | + | + | +) |
| Prunus domestica / insititia | | + | +) | +) | | |
| P. cerasifera | | + | | +) | + | +) |
| P. cerasus | | + | | +) | + | +) |
| Crataegus x media (C. monogyna x C. laevigata) | | | | | | + |
| Sorbus torminalis | | + | | + | + | |
| S. x latifolia (S. torminalis x S. aria) | | + | | | | |
| S. rupicola | | + | | | | + |
| S. lancastriensis | | | | | + | |
| Ribes spicatum | | + | +) | +) | +) | +) |

| taxon | vice-county | 66 | 67 | 68 | 69 | 70 |
|---|---|---|---|---|---|---|
| Epilobium tetragonum / tetragonum | | + | | +) | +) | +) |
| E. tetragonum / lamyi | | +) | | | | |
| E. anagallidifolium | | + | | + | + | + |
| E. hirsutum x E. parviflorum | | | | | +) | |
| E. hirsutum x E. montanum | | +) | | | | |
| E. ciliatum x E. parviflorum | | | + | | | |
| E. obscurum x E. parviflorum | | +) | | | +) | |
| E. obscurum x E. tetragonum | | | | | | +) |
| E. montanum x E. parviflorum | | | | | +) | |
| E. montanum x E. palustre | | | | | | +) |
| E. palustre x E. parviflorum | | | | | | +) |
| E. alsinifolium x E. palustre | | +) | | | +) | +) |
| E. komarovianum | | + | | | | |
| Oenothera biennis | | + | +) | | + | + |
| O. erythrosepala | | + | | + | | |
| Circaea alpina | | | | | + | + |
| C. x intermedia (C. alpina x C. lutetiana) | | +) | + | | + | + |
| Callitriche hermaphroditica | | +) | + | + | + | + |
| C. obtusangula | | | | | + | + |
| C. platycarpa | | + | | | +) | +) |
| Myriophyllum verticillatum | | +) | +) | | +) | +) |
| Bupleurum rotundifolium | | +) | | | +) | +) |
| B. lancifolium | | + | + | | + | + |
| Euphorbia (esula) | | + | +) | +) | + | + |
| E. (uralensis) | | | | +) | | |
| E. cyparissias | | | | +) | +) | +) |
| Polygonum oxyspermum / raii | | +) | + | + | + | + |
| P. rurivagum | | +) | | | | |
| Rumex tenuifolius | | | + | + | | |
| R. x pratensis (R. crispus x R. obtusifolius) | | + | | | | |
| R. x arnottii (R. longifolius x R. obtusifolius) | | + | | | | |
| Ulmus carpinifolia | | + | | | | |
| Alnus incana | | + | +) | | + | +) |
| Populus x canescens | | + | | | | |
| P. x euramericana (P. x canadensis) | | + | | +) | + | + |
| P. candicans | | + | | +) | +) | + |
| P. trichocarpa | | + | | | | + |
| Populus - balsam poplar hybrids | | + | | | | |
| Salix alba x S. pentandra | | | | | + | + |
| S. fragilis x S. pentandra | | + | | | + | + |
| S. triandra x S. viminalis | | | + | +) | | +) |

| taxon vice-county | 66 | 67 | 68 | 69 | 70 |
|---|---|---|---|---|---|
| S. purpurea x S. viminalis | + | +) | +) | +) | + |
| S. caprea x S. viminalis | + | +) | | + | + |
| S. caprea x S. cinerea | + | +) | +) | + | |
| S. caprea x S. myrsinifolia | +) | | | | +) |
| S. caprea x S. phylicifolia | +) | | | | |
| S. x calodendron | + | +) | +) | | |
| S. cinerea x S. viminalis | + | +) | + | + | +) |
| S. cinerea x S. purpurea | +) | | | | |
| S. x forbyana | +) | | + | | +) |
| S. cinerea x S. myrsinifolia | + | + | +) | +) | +) |
| S. cinerea x S. phylicifolia | +) | | +) | +) | +) |
| S. cinerea x S. repens | +) | +) | +) | | |
| S. aurita x S. viminalis | | | | +) | |
| S. aurita x S. viminalis x S. caprea | + | | +) | | |
| S. aurita x S. purpurea | | | + | | |
| S. aurita x S. caprea | +) | | | +) | +) |
| S. aurita x S. caprea x S. cinerea | | | | | +) |
| S. aurita x S. cinerea | + | | + | +) | +) |
| S. aurita x S. myrsinifolia | | | | +) | +) |
| S. aurita x S. phylicifolia | +) | | | +) | |
| S. aurita x S. repens | +) | | | +) | +) |
| S. myrsinifolia x S. phylicifolia | + | + | + | +) | +) |
| S. myrsinifolia x S. repens | + | | | | |
| S. phylicifolia x S. purpurea | | + | | | |
| S. phylicifolia x S. repens | +) | | +) | | |
| Pyrola media | +) | + | +) | +) | +) |
| Lysimachia terrestris | | | | + | |
| L. ciliata | | | | + | +) |
| L. punctata | + | +) | +) | + | + |
| Anagallis arvensis f. azurea | +) | | | | |
| A. foemina | + | +) | +) | +) | +) |
| Centaurium littorale | | | + | + | + |
| C. capitatum | | | + | | |
| Symphytum orientale | + | + | + | | |
| S. ibiricum | + | + | | | |
| Calystegia sepium / roseata | | | | | + |
| Linaria x sepium (L. repens x L. vulgaris) | + | + | | + | + |
| Mimulus guttatus x M. luteus x M. cupreus | + | | | | |
| M. luteus | + | | | | |
| Veronica x lackschewitzii (V. anagallis-aquatica x V. catenata) | + | | +) | +) | +) |
| Euphrasia arctica / borealis | + | + | + | + | +) |

| taxon | vice-county | 66 | 67 | 68 | 69 | 70 |
|---|---|---|---|---|---|---|
| E. rivularis | | | | | + | + |
| E. anglica | | | | | +) | |
| E. rostkoviana / rostkoviana | | | + | + | + | +) |
| E. rostkoviana / montana | | + | + | + | + | +) |
| E. tetraquetra | | | + | + | +) | + |
| E. scottica | | +) | + | + | + | + |
| E. frigida | | | | + | + | + |
| E. confusa x E. nemorosa | | + | | | + | + |
| E. confusa x E. scottica | | + | | + | + | + |
| E. micrantha x E. nemorosa | | +) | +) | | | |
| E. arctica x E. nemorosa | | ?+) | | | | |
| E. confusa x E. rostkoviana | | | | | | ?+) |
| E. rivularis x E. rostkoviana / rostkoviana | | | | | | ?+) |
| E. rivularis x E. rostkoviana /montana | | | | | | ?+) |
| Utricularia vulgaris, sensu lato | | + | +) | +) | + | + |
| U. intermedia | | +) | +) | | + | + |
| Pulegium vulgare | | | | | +) | +) |
| Mentha x gentilis (M. arvensis x M. spicata) | | + | + | + | +) | + |
| M. x smithiana (M. aquatica x M. arvensis x M.spicata) | | + | | | +) | |
| M. x verticillata (M. aquatica x M. arvensis) | | + | + | | + | + |
| M. suaveolens | | + | | + | | |
| Lamium moluccellifolium | | + | +) | + | | |
| L. hybridum | | + | +) | + | + | + |
| Galium x pomeranicum (G. album x G. verum) | | + | + | | + | + |
| Sambucus canadensis | | + | | | | |
| Valerianella locusta / dunensis | | | | | + | + |
| Galinsoga parviflora | | + | | | | |
| G. quadriradiata | | + | | | | + |
| Senecio vulgaris f. radiatus | | + | | | + | |
| S. x londinensis (S. squalidus x S. viscosus) | | + | | | | |
| S. x ostenfeldii (S. aquaticus x S. jacobaea) | | + | | | + | + |
| Petasites hybridus - female | | + | + | | | + |
| Aster x salignus (A. novi-belgii x A. lanceolatus) | | + | | | + | + |
| Anthemis arvensis | | + | +) | + | +) | +) |
| Chamaemelum nobile | | +) | | | +) | +) |
| Artemisia verlotiorum | | + | + | | | |
| Arctium minus | | + | | | | ?+ |
| Hypochoeris glabra | | +) | | | | + |
| H. maculata | | +) | | | + | |
| Lactuca virosa | | + | +) | + | + | +) |
| Hieracium pilosella / micradenium | | + | | | | |

| taxon | 66 | 67 | 68 | 69 | 70 |
|---|---|---|---|---|---|
| H. pilosella / euronotum | + | | | | |
| H. pilosella / pilosella | + | | | + | |
| H. pilosella / trichosoma | + | | | | |
| H. pilosella / tricholepium | + | | | + | |
| H. pilosella / melanops | + | | | | |
| H. pilosella / trichoscapum | + | | | | |
| Crepis setosa | | + | + | + | |
| C. vesicaria / taraxacifolia | + | +) | +) | | |
| C. biennis | + | + | +) | + | + |
| C. foetida | +) | +) | | | |
| Taraxacum Sect. Platyglossum | + | | + | | |
| Alisma lanceolatum | ?+) | | | | |
| Hydrilla verticillata | | | | + | |
| Zostera marina | +) | | +) | + | +) |
| Z. angustifolia | +) | + | + | +) | |
| Z. noltii | +) | + | + | | |
| Potamogeton coloratus | +) | +) | + | + | +) |
| P. alpinus | + | + | +) | + | + |
| P. lucens | +) | + | + | +) | +) |
| P. praelongus | +) | + | +) | +) | + |
| P. friesii | +) | +) | | | |
| P. acutifolius | | +) | | | |
| P. filiformis | | + | | | |
| P. x zizii (P. gramineus x P. lucens) | | + | | +) | +) |
| P. x nitens (P. gramineus x P. perfoliatus) | | + | +) | +) | + |
| P. x cooperi (P. crispus x P. perfoliatus) | +) | | | | |
| P. x lintonii (P. crispus x P. freisii) | | + | + | | |
| P. x salicifolius (P. lucens x P. perfoliatus) | | | + | | |
| P. x olivaceus (P. alpinus x P. crispus) | | | + | | + |
| P. suecicus (P. filiformis x P. pectinatus) | | | + | | |
| Ruppia cirrhosa | +) | | | | |
| R. maritima | + | +) | + | + | + |
| Najas flexilis | | | | + | |
| Allium scorodoprasum | + | | | + | + |
| A. schoenoprasum | + | + | | + | +) |
| A. carinatum | + | | + | | |
| A. triquetrum | + | | | +) | |
| A. paradoxum | + | | + | | +) |
| Juncus compressus | + | | | +) | + |
| J. alpinus | + | | | + | |
| J. x buchenaui (J. alpinus x J. articulatus) | + | | | | |
| J. x surrejanus (J. acutiflorus x J. articulatus) | | | | | + |

| taxon vice-county | 66 | 67 | 68 | 69 | 70 |
|---|---|---|---|---|---|
| J. foliosus | | | | | + |
| J. ambiguus | | | | + | + |
| J. x diffusus (J. effusus x J. inflexus) | + | +) | +) | +) | |
| Iris versicolor | | | | + | + |
| I. versicolor x I. virginica | | | | + | |
| Epipactis phyllanthes | | + | | + | |
| E. atrorubens | + | | | + | |
| E. dunensis | | | + | + | |
| E. leptochila | | + | | | |
| x Dactylogymnadenia cookei (Dactylorhiza fuchsii x Gymnadenia conopsea) | +) | | | +) | |
| x D. varia (Dactylorhiza purpurella x Gymnadenia conopsea) | +) | | | | |
| x D. legrandiana (Dactylorhiza maculata x Gymnadenia conopsea) | +) | | | | |
| Dactylorhiza x transiens (D. fuchsii x D. maculata) | +) | | | + | |
| D. x kernerorum (D. fuchsii x D. incarnata) | + | | | +) | + |
| D. x venusta (D. fuchsii x D. purpurella) | + | | | + | + |
| D x grandis (D. fuchsii x D. praetermissa) | + | | | | |
| D. x claudiopolitana (D. incarnata x D. maculata) | +) | | | | |
| D. x formosa (D. maculata x D. purpurella) | +) | +) | | +) | + |
| D. x wintoni (D. incarnata x D. praetermissa) | +) | | | | |
| D. x latirella (D. incarnata x D. purpurella) | +) | | | | |
| D. praetermissa | + | | | | |
| Sparganium angustifolium | + | | | + | + |
| S. minimum | | + | +) | + | + |
| Scirpus lacustris / tabernaemontani | + | + | + | + | +) |
| S. cernuus | | | | | + |
| Eleocharis acicularis | | + | +) | + | + |
| E. austriaca | | + | | | + |
| Carex flava | | | | + | |
| C. flava x C. demissa | | | | + | |
| C. serotina | | + | + | + | + |
| C. limosa | | + | +) | + | + |
| C. magellanica | | + | | + | + |
| C. ericetorum | + | | | + | |
| C. ornithopoda | | | | + | |
| C. digitata | | | | + | |
| C. elata | +) | | | + | + |
| C. aquatilis | +) | + | | + | + |
| C. x boenninghausiana (C. paniculata x C. remota) | | | + | + | +) |
| C. x pseudoaxillaris (C. otrubae x C. remota) | | | + | | |
| C. spicata | + | + | | + | + |

| taxon vice-county | 66 | 67 | 68 | 69 | 70 |
|---|---|---|---|---|---|
| C. muricata / lamprocarpa | | + | + | + | + |
| C. divulsa / leersii | + | | | | |
| C. demissa x C. hostiana | | | | + | |
| C. lepidocarpa x C. hostiana | + | + | | | |
| C. x turfosa (C. elata x C. nigra) | | + | | + | |
| Glyceria x pedicellata (G. fluitans x G. plicata) | + | +) | +) | + | + |
| Festuca tenuifolia | + | + | + | +) | +) |
| F. longifolia | +) | | +) | | |
| F. juncifolia | + | | | | |
| x Festulolium loliaceum (Festuca pratensis x Lolium perenne) | + | +) | +) | + | + |
| x F. braunii (Festuca pratensis x Lolium multiflorum) | + | | | | |
| x F. frederici (Festuca rubra x Lolium perenne) | | | | | + |
| Poa angustifolia | + | | | | |
| Bromus lepidus | + | | | +) | +) |
| B. hordeaceus / thominii | + | +) | +) | | |
| B. x pseudothominii (B. hordeaceus x B. lepidus) | + | + | +) | | |
| B. secalinus | +) | +) | +) | | |
| B. arvensis | +) | | | +) | +) |
| B. racemosus | | +) | +) | +) | +) |
| B. commutatus | + | +) | +) | + | +) |
| Elymus x obtusiusculum (E. pycnanthus x E. junceiforme) | + | +) | + | | +) |
| E. x laxum (E. junceiforme x E. repens) | + | | | | |

\+ recorded recently in the vice-county

+) recorded in the vice-county, but not recently

Alphabetical Index